The
Humble
Approach

The
Humble
Approach

Scientists Discover God

JOHN MARKS TEMPLETON

TEMPLETON FOUNDATION PRESS
Philadelphia & London

Templeton Foundation Press
Five Radnor Corporate Center, Suite 120
100 Matsonford Road
Radnor, Pennsylvania 19087

First cloth edition published in 1981 by William Collins Sons & Co Ltd
Second cloth edition published in 1981 by The Seabury Press
Revised paperback edition published in 1995 by Continuum
Templeton Foundation Press paperback, April 1998

Printed in the United States of America

Library of Congress Cataloging-in-Publication Data

Templeton, John, 1912–
The humble approach: scientists discover God/ John M.
     Templeton.
          p.   cm.
     Includes bibliographical references (p.   ).
     ISBN 1-890151-17-3
     1. Science and religion.   2. Humility.   I. Title.
BL240.2.T433   1994
210—dc20   94-36050
                                                      CIP

CONTENTS

Appendixes

I   An Introduction

We are perched on the frontiers of future knowledge. Even though we stand upon the enormous mountain of information collected over the last five centuries of scientific progress, we have only fleeting glimpses of the future. To a large extent, the future lies before us like a vast wilderness of unexplored reality. The God who created and sustained His evolving universe through eons of progress and development has not placed our generation at the tag end of the creative process. He has placed us at a new beginning. We are here for the future.

Our role is crucial. As human beings we are endowed with mind and spirit. We can think, imagine, and dream. We can search for future trends through the rich diversity of human thought. God permits us in some ways to be co-creators with Him in His continuing act of creation.

There is, however, a stumbling block: egotism. The closed-minded attitude of those who think they know it all inhibits future progress. Natural scientists, by and large, have overcome this hurdle. They are more open-minded. They research the natural wonders of the universe, devising new hypotheses, testing them, challenging old assumptions, competing with each other in professional rivalry. The physical future of human civilization is in their professional hands, guided by relatively tolerant and open minds.

This is not equally true concerning our spiritual future.

Some theologians, religious leaders, and lay people are frequently blind to the obstacles they themselves erect. Many are not even aware that the spiritual future could, or should, be different from anything that has ever been before. Many do not realize that spiritual reality can be researched in ways similar to those used by natural scientists. Some do not want even to consider the possibility of a future of progressively unfolding spiritual discoveries.

Why not? Many devoutly religious people are not devoutly humble. They do not admit their worldview is limited. They are not open to suggestions that their personal theology might be incomplete. They do not entertain the notion that other religious have valuable insights to contribute to an understanding of God and His creation. When people take a more humble attitude, they welcome new ideas about the spirit just as they welcome new scientific ideas about how to cure headaches, how to heat and cool their homes, or how to develop natural resources.

The humble approach to human knowledge is meant to help as a corrective to the parochialism that blocks further development in religious studies. Taking this humble approach reminds us that each person's concept of God, the universe, even his or her own self is too limited. To some extent, we are all too self-centered. We overestimate the small amount of knowledge we possess. To be humble means to admit the infinity of creation and to search one's place in God's infinite plan for creation. This approach asks each of us, whether we are students of the natural or the supernatural, to witness to the intimate relationship of physical and spiritual reality in our own lives. In a humble manner we can use our talents to explore the universe to discover future trends. There is abundant evidence that by grace God gives us talents and intelligence with which to participate in His vast creative process.

Until the emergence of human beings on the face of the earth, evolution proceeded routinely, unfolding the rich complexity of mineral, vegetable, and animal life. Now with human intelligence capable of studying the Creator and His creation, evolution no longer travels only on its own path. Possibly it was

God's plan that one day His children could serve as useful tools for His creative purposes.

At this present moment, the human race, even after thousands of years of historical development, is still at the dawn of a new creation. This is a tremendous, awe-inspiring responsibility. It should humble us.

In fact, humility is the key to progress. Without it we will be too self-satisfied with past glories to launch boldly into the challenges ahead. Without humility we will not be wide-eyed and open-minded enough to discover new areas for research. If we are not as humble as children, we may be unable to admit mistakes, seek advice, and try again. The humble approach is for all of us who are concerned about the future of our civilization and the role we are to play in it. It is an approach for all of us who are not satisfied to let things drift and who want to channel our creative restlessness toward helping to build the kingdom of God.

Every person's concept of God is too small. Through humility we can begin to get into true perspective the infinity of God. This is the humble approach.

It is also in humility that we learn from each other, for it makes us open to each other and ready to see things from the other's point of view and share ours with him freely. It is by humility that we avoid the sins of pride and intolerance and avoid all religious strife. Humility opens the door to the realms of the spirit, and to research and progress in religion.

Twenty-five centuries ago Xenophanes and twelve centuries ago Shankara taught that nothing exists independently of God and that God is immeasurably greater than all time and space, let alone the visible earth. But only in the last century have modern sciences come to realize how very tiny the earth is compared to the billions of galaxies and how very brief human history is in the ongoing creation of the universe. Only now is man beginning to find evidence of hitherto undreamed-of forces and even dimensions of reality which transcend the invisible space-time field which holds together within its astonishing configuration all that man can observe in space and time. There are vast realms reaching out beyond the known, inspiring wonder and inviting inquiry, but it is humility that opens the way forward.

It is to be hoped that this book will reach many people who are ready to benefit from this kind of experience at the frontiers of knowledge where their minds may be stretched far beyond the range of their grasp hitherto. Perhaps people will be uplifted and inspired by catching glimpses of unexpected aspects of reality that beckon their inquiry with exciting promises of still further manifestations of truth beyond anything they could anticipate. Perhaps the advance into the realms of spiritual reality and progress in religion will be as outstanding and rapid as the astonishing advances in physics, astronomy, and genetics. Church denominations may then be inspired to devote the manpower and find the funds needed for the promotion of research. Young people may be attracted to a religion that is genuinely dynamic and rapidly progressing. The visions and the teachings of the great prophets of the past need not be discarded or disputed. Rather they should be studied again and used as springboards to new and greater understanding and love of God. This book explores the possibility that humility in man's understanding of God may be more fruitful than the formal systems of thought which we have inherited, whether they be theistic, pantheistic, or panentheistic. Gradually we may learn to love every one of God's children and be grateful for an increasingly rich diversity of thought emanating from research and worship in every land. One of the purposes of this book is to examine and foster the idea that through a humble approach in knowledge in which we are open-minded and willing to experiment, theology may produce positive results even more amazing than the discoveries of scientists which have electrified the world in this last century.

Why do millions of people think theology has become obsolete, when no one thinks physics or astronomy can become obsolete? That is the subject of this book, *The Humble Approach*. Theology was called the queen of the sciences in ages gone by and can deserve that title again when it adopts the humble approach.

A bibliography on the subject of science and religion

is included to encourage others to read and write more extensively in this important and developing field. The bibliographies include articles and books for beginners as well as for advanced thinkers, whether they are scientists or theologians. By reading and writing in this theological field, scientists and other laymen may not only enhance their own spiritual growth but also stimulate progress and expand the whole field of theology in ways that may benefit all. Let us hope that already a spiritual and religious renaissance may have started, and that a great new day may be dawning.

II The Blossoming Time of Man

We should be overwhelmingly grateful to have been born in this century. The slow progress of prehistoric ages is over, and centuries of human enterprise are now miraculously bursting into flower. The evolution of human knowledge is accelerating, and we are reaping the fruits of generations of scientific thought: More than half of the scientists who ever lived are alive today. More than half of the discoveries in the natural sciences have been made in this century. More than half of the goods produced since the earth was born have been produced in the twentieth century. Over half the books ever written were written in the last half-century. More new books are published each month than were written in the entire historical period before the birth of Columbus.

Many astronomers believe the universe began with a "Big Bang" about eighteen billion years ago. This vast figure is about the same as the number of minutes which have elapsed since Moses was born. But not until five billion years ago did our galaxy, the Milky Way, containing over one hundred billion stars, look as it does now. At about that time, our star, the sun, was formed. About five billion minutes ago, William the Conqueror was born in Normandy. Thus seventy-two percent of the history of the universe occurred while, to quote from

Genesis, "The earth was without form and void and darkness was on the face of the deep; and the Spirit of God moved upon the face of the waters."

According to most astronomers, the earth was formed about four and one-half billion years ago. Dr. Elso Barghoorn, a paleontologist at Harvard, believes he has found evidence in Swaziland that the first living things appeared on earth three and two-fifth billion years ago. These were single-cell plants resembling algae. Three and two-fifth billion minutes ago, the year was 1330. The scourge of the Black Death had begun to spread throughout Europe and ultimately killed a fourth of the total population of the continent. It is interesting to note that while single-cell creatures first appeared on earth over three billion years ago, it has taken eighty percent of the total period of evolution since that time to evolve a creature with *more* than one cell. Then microbes slowly evolved into worms, fishes, reptiles, and mammals. Humans did not appear until forty million years ago. For comparison, forty million minutes is equivalent to only seventy-six years.

It has been only six thousand years since man invented any form of writing. Thus, only the last one-seventeenth of one percent of man's existence has been within the age of written communication; and it has only been this age that has allowed the full flowering of the human intellect. For comparison, six thousand minutes is only four days.

Even then, progress continued to be slow. Ninety-seven percent of recorded history passed by before man began his rapid discovery of the secrets of nature. Two hundred years ago, there was no faster transportation than the horse on land and sailing ships on the seas. Communications two hundred years ago were scarcely faster than they had been in the days of Moses. Energy still came mostly from muscle power. Electricity was a laboratory curiosity. Germs were unknown. Photography was unheard of. Less than a million people on earth could read or write. Ninety-five percent of the workers in the world had jobs in agriculture and fishing, compared with only four percent in the United States today.

It was only about a century ago that James Clerk Maxwell

proved that light, electricity, and magnetism are all part of one continuous spectrum of wave lengths. Then followed the great discoveries of radio waves, X rays, microwaves, infrared and ultraviolet light, and with them the invention of radar, radio astronomy, television, and lasers. All of this had to do with particles of electromagnetic energy traveling at the speed of light.

After Maxwell's theory, three other major developments in physics occurred in this century: first Einstein's theory of relativity; then quantum mechanics; and, finally, antimatter. The great science of nuclear physics is less than a century old. The electron was discovered in 1914 by Ernest Rutherford, and the neutron in 1932 by James Chadwick. Less than fifty years ago physicists thought there were only three kinds of subatomic particles: electrons, protons, and neutrons. Today so many new particles are being discovered through sophisticated technology that Lawrence Berkeley Laboratory began to issue a new list of subatomic particles twice yearly.

Less than a century ago the only cosmic forces known were those of gravity and electromagnetism. Then the Strong Force and the Weak Force were discovered. The Strong Force binds protons and neutrons together in the nuclei of atoms. It is 10^{40} times stronger than gravity but operates only over the tiny distances inside the nucleus. The Weak Force operates in some processes where atomic particles are transformed. Since 1970 physicists have found two more cosmic forces, the Color Force and Weak Force Mark II.

Astronomy, the oldest science, has been revolutionized in the last fifty years. Edwin Powell Hubble and others proved for the first time in the 1920s that there are other galaxies far beyond our Milky Way. Today we believe there are over one hundred billion such galaxies. Sir Bernard Lovell helped initiate the science of radio astronomy in 1943. More recent is the science of X-ray astronomy. Quite recently astronomers have discovered variable stars, pulsars, and quasars. Since 1960 when Allan Sandage and others collected evidence about quasars, astronomers have estimated that there are over fifteen million quasars in the universe, the nearest being over a billion light years away, and others as far distant as six billion light years. It was not until 1968 that pulsars were discovered, and now astronomers think there may be over one hundred thousand of them in our Milky Way galaxy alone. In the past it appeared that the black distance between stars and galaxies was empty, but evidence now indicates that over half the matter in the universe is in those seemingly empty spaces. Some astronomers think there are also millions of black holes whose gravity is so great that no light can escape from them.

In 1911 Victor Hess proved that we are bombarded constantly from all directions by cosmic rays, rays which are electrically charged atomic nuclei that may have originated in the millions of supernovas that have exploded in our galaxy since it began. In 1931 Karl Jansky, a radio engineer, discovered that microwaves from the sky continuously surround and penetrate us. Microwaves travel in straight lines unaffected by magnetic fields. They are the shortest of the radio waves, but a million times as long as light waves. Only since 1950 have scientists been able to detect, with particle accelerators, the neutrinos produced when protons turn into neutrons through the nuclear reactions in the sun's core. Unnoticed by us, these unseen particles rain down on us by day; and even at night, when the earth is between us and the sun, neutrinos pass through the earth in quantities not measurably diminished. Until recently we would have denied the existence of all these unseen forces.

Things are not what they seem. Sometimes phenomena which appear real to us are actually hoaxes perpetrated by our lack of knowledge and limited senses. For example, until five hundred years ago it was assumed that lying in bed was a relatively motionless experience. This seemed an obvious fact to anyone who had ever done it. But Copernicus' discovery that the earth and the planets move around the sun implied that because the earth rotates, a person sleeping in bed moves eastward at one thousand miles an hour. The sleeper also flies one thousand and eighty miles a minute in another direction due to the earth's revolution around the sun. Just a few years ago the rotation of the Milky Way was measured, indicating that our solar system is moving at one hundred and sixty-two miles per second in yet a third direction. Also, in 1977, Ames Research

Center in California computed that our galaxy is speeding away from the original point of the big bang at four hundred miles per second toward a spot in the sky near the Constellation Hydra. So a sleeper may *seem* to be motionless, but in reality he or she has traveled a distance greater than that to the farthest point on earth, in more than four directions at once, and in less time than it took to read this page.

In humility we should admit that other universes may exist unknown to us, and not only in three dimensions but maybe in four, five, or six dimensions. Nigel Calder has argued that other universes are unobservable because, if we *could* observe them, they would be part of our own universe. But if we do not define "exist" too strictly, we may posit the existence of other universes of space and time. He suggested that, "There have been serious speculations about a co-existent anti-universe, or other realms beyond the horizon where the laws of nature were not the same. Is our universe a black hole in somebody else's universe?"[1]

Sir Bernard Lovell wrote that we have "never obtained scientific answers to the problem of whether the universe is finite or infinite, or how it began, and how it will end. Nevertheless, the perspective of these questions and the nature of the possible scientific answers have constantly changed since ancient times."[2]

This should be enough to cause all men and women to pause humbly before the majesty and infinity of what Jefferson called, "nature and nature's God." Discovery and invention have not stopped or even slowed down. Who can imagine what will be discovered if this acceleration continues? Now even the acceleration of discovery seems to be accelerating. The more we learn about the universe the more humble we should be, realizing how ignorant we have been in the past and how much more there is still to discover.

This is the Humble Approach: to assume a realistic attitude before the Creator and admit that we are not the center of the universe. The sun does not revolve around us. Our five human senses are able to comprehend only a small portion of the mysteries, forces, and spiritual realities surrounding us.

Egotism has been a major cause of many mistaken notions in the past. Egotism caused men to think that the stars and the sun revolved around them. Egotism caused men to think that mankind was as old as the universe. Egotism is still our worst enemy. In fact, things are still not what they seem. Only by becoming humble can we learn more. Forces still undreamed of are probably present around us and in us. And more revelations about God's universe will probably be discovered in the next century than in all the millenniums before. Those who believe only what they see are hopelessly self-centered and lacking in humility.

A classic example of how self-centered egotism can block scientific progress is the case of the heliocentric concept of the universe. As Lovell explains, the concept that the sun was the center of the universe afforded "centuries of a comforting philosophical stability in which the Universe was envisaged as made for many by God, who had in the beginning endowed the particles of the Universe with such properties as were necessary for the appropriate inertial motions and gravitational forces." It would not have taken two centuries to challenge this notion, he believes, were it not for the "egocentric conviction that man must be near the center of the universe. Indeed, a retrospective judgment is that the abandonment of the idea of the Sun-centered universe would not have been delayed until the twentieth century but for this egoism of man." In this case, the scientific problem of measuring the distance between stars was overcome more easily than the psychological and spiritual problem of human self-centeredness.

Humility is the gateway of knowledge. To learn more, we must first realize how little we already know. The unknown before us may be a million times greater than what we now know, despite the myriad discoveries made in recent years. Even scientists on the cutting edge of new theories about the universe admit this. Ultimate reality is vastly greater than the sum of phenomena already observed. More and more, the immensity of the physical universe points to a nonphysical Creator who is infinite. The president of a science association said recently that if a fair sample of natural scientists had been asked fifty years ago whether they believed in God, possibly twenty percent would have said yes, whereas today that figure might be as high as eighty percent. The scientists of the 1980s, in other words, believe in the unseen to a far greater extent than did scientists of the 1880s. The modern scientist is more humble than his predecessors.

By taking the humble approach, scientists are acknowledging that in spite of tremendous scientific breakthroughs in recent times, there is still an infinite amount of knowledge to be learned. Many accept an infinite God as the originator of this infinite body of knowledge. In the words of Thomas Carlyle a century ago, they might say:

All visible things are emblems. What thou seest is not there on its own account; strictly speaking it is not there at all. Matter exists only spiritually, and to represent some idea and body it forth.[3]

The question before us is whether theologians and religious scholars, clergy and laity, are also taking the humble approach. Are they affirming the infinite which surrounds us? If they accept the inexhaustibility of God's revelation in terms of science, as do many scientists, they ought to admit that God's revelations in terms of the spirit are also inexhaustible, vastly exceeding our capacities to grasp them. The greater part of divine revelation, both scientific and spiritual, may still be ahead of us, not behind us.

For some more recent thoughts on "reality," see our book *Is God the Only Reality?* (New York: Continuum, 1994).

III The Vast Unseen

Sir James Jeans, looking at the millions of other galaxies, said:

Do their colossal incomprehending masses come nearer to representing the main ultimate reality of the universe, or do we? Are we merely part of the same picture as they or is it possible that we are part of the artist? Are they perchance only a dream, while we are brain cells in the mind of the dreamer?

Each year more leading scientists express their belief in God. Some say that nothing exists except God. Dr. Allan Sandage, director of Mount Wilson Observatory, explains:

The world is incredible—just the fact that you and I are here, that the atoms of our bodies were once part of stars. They said I am on some kind of religious quest, looking for God, but God is the way it's put together. God is Newton's and Einstein's Laws.

Sandage is not alone in his quest for God. Other scientists accompany him.

In his highly acclaimed introduction to Einstein, Lincoln Barnett wrote in 1957:

In the evolution of scientific thought, one fact has become impressively clear: there is no mystery of the physical world which does not point to a mystery beyond itself . . . . Man's inescapable impasse is that he himself is part of the world he seeks to explore; his body and proud brain are mosaics of the same elemental particles that compose the dark drifting clouds of interstellar space; he is, in the final analysis, merely an ephemeral conformation of the primordial space-time field. Standing midway between macrocosm and microcosm, he finds barriers on every side and can perhaps but marvel, as St. Paul did nineteen hundred years ago, that the world was created by the word of God so that what is seen was made out of things which do not appear.[4]

The famous Cambridge astronomer, Sir Arthur Eddington, tried to unite quantum physics and relativity with what he called his own "mysticism," his conviction that the universe worth studying is the one within us. He suggested that man should use "the higher faculties of his nature, so that they are no longer blind alleys but open out into a spiritual world—a world partly of illusion, no doubt, but in which he lives no less in the world, also of illusion, revealed by the senses."

Eddington wrote a parable about the small-minded egotism of a marine biologist who cast his net into the sea to accumulate a mass of evidence on what dwells in the deep. From the fish in his net, he arrived at two conclusions. First, no creature is less than two inches long; and second, all sea creatures have gills. A critic objected that the sea contained many creatures less than two inches long, but the scientist would have to use a finer net if he wished to catch them. The scornful biologist replied, "Anything uncatchable in my net is by that very fact outside the scope of fish science. What my net can't catch isn't a fish!"

Robert Boyd, a professor of physics at London University, has said,

I think it is quite as common to find Christians among scientists as in any other profession. In fact, I would say that, if anything, it is a little more common to find Christians among physicists than it is in some other branches of science.

In other words, astronomers at work discovering the vast complexities of the macrocosm, and nuclear physicists investigating the awesome variety of the microcosm, are now concluding that the universe could not have happened by chance. The famous physicist Sir James Jeans (1877–1946) said that the universe is beginning to look not like a great machine but like a great thought.

In their own ways, many scientists are reaffirming St. Paul's view that, "Our eyes are fixed not on things that are seen but on the things that are unseen; for what is seen passes away: what is unseen is eternal." (II Corinthians 4:18 NEB). Or as Henry Drummond wrote:

The physical properties of spiritual matter form the alphabet which is put into our hands by God, the study of which, if properly conducted, will enable us more perfectly to read that great book which we call the Universe . . . . Law is great not because the phenomenal world is great, but because these vanishing lines are the avenues into the Eternal Order.[5]

St. Paul would have concurred with Drummond that, "The visible is the ladder up to the invisible; the temporal is but the scaffolding of the eternal."

Ralph Waldo Trine wrote:

Everything exists in the unseen before it is manifested or realised in the seen, and in this sense it is true that the unseen things are the real, while the things that are seen are the unreal. The unseen things are cause; the seen things are the effect. The unseen things are the eternal and the seen things are the changing, the transient.[6]

On a cloudy day there appears to be no sun; but we have faith that it is only hidden. However, if we had lived on Venus which is always totally covered by thick clouds (rather than on earth), agnostics might never have believed that there is really a sun and stars and galaxies.

Lowell Fillmore wrote in 1963, "Until we tune in our mind to perceive God's kingdom, we judge the world by appearances only, and therefore behold only the dark things in the world."

Material things which appear, appear only because God has given us five senses with which to perceive a few traits of a few of the myriad notes in the giant creative symphony of life which surrounds us. The unknown is found to extend vastly beyond the area of the known, even after scientists have multiplied the known a hundredfold as they have in this century alone. Maybe there are many other forms of life unseen by us? There may be other beings occupying the same time and location who have five other senses able to perceive different configurations. In that case, they would not be able to see us anymore than we can see them.

No one yet knows what gravity is. But astronomers calculate that the gravity in the universe is 100 times greater than can be accounted for by visible stars and galaxies. Some guess the vast discrepancy is caused by black holes and others guess thin gas between galaxies. Maybe even other universes inconceivable to us are present in our visible universe.

Within the human body there are as many unseen activities as there are in the world of a beehive. Hundreds of bees die and are replaced by others, but the beehive lives on. Billions of cells work together to produce a human body. Millions die daily and are replaced. Each cell, although invisible to the naked eye, is composed of millions of atoms which in turn are formed by myriad particles and waves.

The physical body has no substance or permanence. It may be only a colony of cells in which a soul may dwell temporarily and develop for a divine purpose, for contact with an external world detected through a complex of sensory nerve endings. It would be appallingly narrow and egotistical to think that nothing exists in the universe except what these "cell colonies" called humans have learned to touch or see. This would be like a worm denying the existence of a butterfly because of its feeble eyesight, or a stone denying its dimensions and weight because it cannot comprehend arithmetic.

Taking the humble approach, one considers our bodies and minds to be simply ever-changing configurations of waves and particles designed by God as temporary habitations for souls in their earthbound years. Our bodies may look opaque to our eyes, but they are as transparent as jellyfish to neutrinos passing through by the millions.

The wind is just as real as flesh and just as temporary. God's chosen way of multiplying flesh is by tiny patterns which fit atoms into molecules (called genes and chromosomes) and molecules into cells. If our body were as dense as some stars, all its subatomic particles would not fill the space of a pin head. Maybe he who thinks of his body and mind as more real than his soul proves he is a pin head! Men are like clouds—formed by the moving spirit, visible only to nerves tuned to certain wavelengths, and temporary.

A tree is a manifestation of God. In a greater sense, man is a manifestation. Likewise, the sense evidence (with which we have to do in observational science) may be also a manifestation of the spiritual world, but only a small part. The seen and the unseen both exist together; but the seen is limited and temporary.

Emanuel Swedenborg wrote that nothing exists separate from God. If God is infinite, then nothing can be separate from Him. In Him we live and move and have our being. God, he claimed, is all of you and you are a little part of God. Swedenborg taught that man is not in heaven until heaven is in man. It is on this material earth that we begin to receive the life and spirit of heaven within us. We are citizens of the spiritual world, and we are spirits from the day of our conception. Love, loyalty, patience, and mercy are more real than are tangible objects, and God seeks to instill these spiritual realities into our own lives here and now. Taking the humble approach one also believes in this natural world as an incubator, provided by God, in which our spirits can develop and seek their ultimate expression in a realm outside and beyond these earthly confines.

To demonstrate the limited ability of the human mind to comprehend the entire reality around us, in England in 1882, a headmaster named Edwin A. Abbott wrote a delightful and humorous fantasy called *Flatland.* His purpose was to help us understand that our minds are very limited so that reality may be incomprehensible to us. The people of Flatland lived in only two dimensions. They denied the possiblity of any being of three dimensions. Yet objects of three dimensions produced miraculous and mysterious phenomena by moving into or out of Flatland. The hero of the story, Mr. Squire, visited the land of no dimension (Spotland) and the land of one dimension (Lineland) whose inhabitants refused to believe in the existence of any creature of two dimensions. After returning from his later visit to the wondrous land of three dimensions (Spaceland) his Flatland rulers put him in prison for life lest he should stir up trouble among Flatlanders by talk of realms which transcend two dimensions. Finally, Mr. Squire says, "I will endure this and worse, if by any means I may arouse in the interiors of Planes and Solid Humanity a spirit of rebellion against the conceit which would limit our dimensions to two or three or any number short of infinity." Mr. Squire begs his readers not to suppose that every minute detail in the daily life of Flatland must correspond to some other detail in Spaceland; and yet he hopes that, taken as a whole, his work may prove suggestive as well as amusing to those Spacelanders of moderate and modest minds who—speaking of that which is of the highest importance, but lies beyond experience—decline to say on the one hand, "This can never be," and on the other hand, "It must needs be precisely thus, and we know all about it."

Like the inhabitants of Flatland, people collect their evidence to prove or disprove the existence of God. Some people have argued that God is our "ultimate end" or the "ground of being" or "original cause" or the "power" that maintains our existence. Philosophers and theologians, living in what we might call "Logicland," have been convinced for centuries that if their understanding of God is grounded in logic, it must be sound. But in humility should we not admit that human logic is much too inadequate to comprehend fully the infinite Creator? Proving that God exists or does not exist might be an endeavor destined to elude the capacities of the human mind no matter how rational it is. We are often like characters in a play trying to prove the existence of the playwright and then guessing his or her name. Or we are like cells in your toe, trying to prove your intelligence.

Some people think supernatural events, such as miracles, are needed to prove God's existence. But natural processes and the laws of nature may be merely methods designed by God for His continuing creative purposes. When new laws are discovered by human scientists, do they not merely discover a little more of God?

Each of us every day is swimming in an ocean of unseen miracles. For example, each living cell is a miracle; and the human body is a vast colony of over a hundred billion cells. The miracle of this body includes both our ability to recognize it as well as our inability ever to exhaust the true significance of it. As Albert Einstein said, "The most incomprehensible thing about the universe is that it is comprehensible."[7] That the universe exhibits order, not chaos, suggests the futility of trying to fathom the nature of matter without investigating the unseen spirit behind it. Each time new laws are discovered by scientists, however, we learn a little more about God and the ways He continually maintains and is building His creation.

It is reported that William Jennings Bryan squelched the famous atheist Ralph Ingersoll on this very point of recognizing the unseen in the seen. One day, Ingersoll saw an attractive globe of the earth in Bryan's office and asked, "Who made it?" Bryan replied, "No one. It just made itself." Similarly, those who argue that the universe is governed only by chance would do well to investigate the scientific studies of purposeful activity from the most simple one-cell organism responding to external stimuli and trying to maintain homeostasis, to the human animal who plans long-range goals and contemplates life's ultimate purpose. Would it not be strange if a universe without purpose accidentally created humans who are so obsessed with purpose?

A mythical observer from another universe, who might have witnessed the spectacular "Big Bang" when our universe was created about eighteen billion years ago, would have seen after the first year only a vast blackness with thin clouds of stars and other fragments flying apart. But we, who observe from the surface of our small planet Earth, see a totally different picture. We see a drama of evolution and progress on the surface of our earth which is truly amazing and miraculous. And this progress is speeding up faster and faster and faster. By an unbelievable miracle, billions of humans, each of whom is a colony of billions of atoms, have suddenly covered the face of the earth. Most amazing of all is the fact that the unseen minds of these humans are accumulating knowledge in explosive proportions—knowledge of themselves, of the universe, of their Creator. Could we ever make that observer from another universe believe this unseen explosion of human knowledge really exists? Would he believe that these new invisible minds are themselves participating creators in the ongoing drama of evolutionary creation? Or would he say that spaceships now traveling from earth just happened by chance?

The extent to which spirit and matter, the unseen and the seen, are related has been hotly debated over the centuries by theists, atheists, agnostics, pantheists, and others. The fact that these arguments continue suggests that this is still a lively question among thinking people. The vast unseen has not lost its power to tantalize our imaginations.

Traditionally, theologians have conceived God as the transcendent Lord, utterly differentiated from His creations, and sometimes as quite detached from them, but how can He be the infinite Creator and yet be detached from His creations, for He is the continuous source of their existence? Does not belief in God as Creator mean that He does not exist for Himself alone but for all that He has made and continues to support in being, through the embrace of His presence and love and power?

For centuries philosophers have debated the concept of pantheism, which means that God is everything and everything is God, and that his Spirit is present in every creature and thing. Primitive pantheists held the animistic belief that there were spirits in plants, animals, landscapes, and idols. Pantheism conceives of the Spirit dwelling in a creature as an emanation of God's being. Some modern thinkers along this line reject the idea that two separate realities dwell together in one creature but rather that nothing exists in separation from God. Paul Tillich, a Christian theologian, sometimes referred to God as the ground of being. But Mary Baker Eddy, Charles Fillmore, and Ernest Holmes went further to suggest that matter may be only an outward manifestation of divine thought, and that the creative spirit called God is the only reality.

The concept that each thing or being has its reality in God, however, contrasts sharply with the idea that material things are mere illusions not worth studying, or the pantheist notion that God is identical with nature and restricted to visible, material things. Both concepts contrast sharply with the humble approach where one takes the position that God is infinite, whereas created objects are contingent and finite. Even the vast universe falls infinitely short of what God is. God is continually creating each object and person just as he is creating his universe. For thousands of years men and women were so short-sighted and self-centered that they thought the universe was not much larger than the earth. With more humility and knowledge, we now think the universe is a billion, billion, billion times more massive. Thus, we realize that the infinite God is even "more infinite" (if we may coin a phrase) than we had supposed.

If God is infinite, then it follows that all other reality is dependent on Him and cannot exist apart from Him. Matter and energy may be only contingent manifestations of God. Space and time may be only manifestations of God. We should not think of matter and energy as created by God but as now utterly independent of God. That would mean that God is not "all in all" the creative Ground and Sustainer of all that is. Matter and energy may be only creaturely manifestations of the universal Creator. While God does not need the universe to be God, the universe may need to be unceasingly supported and enfolded in His presence and power to be what it is. Maybe it can only exist in and through God.

The excitement and importance of scientific study of nature and the cosmos are enhanced (not reduced) if we conceive of each discovery as a new revelation of a reality deriving from and grounded in God.

When new discoveries point more to the nonexistence of matter, it becomes easier to think of matter and spirit as a unity. Einstein's theory of relativity makes it easier to understand that space and time may not be exactly what they appear to be. New discoveries cause us to be increasingly humble about saying that what we see is real and what we do not see is illusion or myth. Lowell Fillmore said, "Remember that although God's principles are spirit and cannot be seen, they are more real than tangible things. God's principles are fundamental, and infinitely more far-reaching than the principles of mathematics. His invisible principles uphold that which is visible to us." Or, as it is stated in the letter to the Hebrews, "By faith we understand that the world was created by the word of God, so that what is seen is made out of things which do not appear." (Hebrews 11 : 3 RSV.)

Traditional pantheism serves a good purpose in suggesting the close intimacy of spirit and matter and the personal relationship between the Creator and creation. But it is not compatible with the Christian concept of a personal God who loves all of us, and numbers the hairs of our heads. The profound mutual indwelling between man and God may be better stated by the Unity School of Christianity, "God is all of me: and I am a little part of him." Such a notion implies an interdependent relationship between God and us. As even "a little part of him," we realize the mutual unity of God and his creation. We realize that our own divinity arises from something more profound than merely being "God's children" or being "made in his image."

To the atheist, the tiny fraction of creation that can be seen and analyzed is the only reality. The confirmed atheist does not believe in either a God who permeates this creation or in one who is transcendent and totally separate from created matter. The humble men and women, however, acknowledge that we know very little about the myriad forces and beings around us and that what both the ignorant and wise call facts may be no more than optical illusions. Those choosing the humble approach shun the colossal conceit of the atheist who believes that the material world is the only reality and that it is totally explicable without a creator.

Like Thoreau, humble people hope that some ponds will always be thought of as bottomless so that the concepts of the infinite and the unseen and the unfathomable will be part of our daily experience. Would Karl Marx, Thoreau's contemporary, have dismissed as superstition the idea that his room was full of invisible cosmic rays because the scientific knowledge and technology of his day could not detect or explain them? As an atheist, he should have believed only in the material world he could perceive. How unfortunate to have been an atheist living before the invention of technology that could appreciably extend the information acquired by man's five limited senses. Who today does not have faith in cosmic rays and radio waves, even though they are invisible? As a child of the nineteenth century, Marx believed that his kitchen table was composed of solid matter. Now even scientists living in formerly Marxist nations believe a table is, to a large degree, space or nonmatter. Its solid appearance comes only from responses by man's senses to billions of invisible electrons and wavelike particles vibrating at the speed of light in various patterns.

In 1848 when Marx was denying that people had immortal souls, psychosomatic medicine was considered superstition. Today it is widely believed that health can be restored to the body not only by chemicals but by mind and spirit as well. Recent studies around the world reveal the close interaction of body and mind, the mortal and the immortal, in the healing process. More and more doctors are now talking about "healing the patient" in addition to "curing the disease." The latter may be achieved by chemistry and physical therapy. The former, however, requires psychic and spiritual remedies not always available to the health care delivery team. Total health is not reducible to material explanations alone.

Agnostics sometimes claim that atheists who soundly deny God are small-minded, but they themselves are open-minded in affirming their uncertainty in saying they do not know. But many have erected an impenetrable barrier of doubt around their minds. This type of agnostic is like a blind man who will not admit that a rainbow is beautiful or that others enjoy it because he cannot see it personally. The more committed agnostics are to their doubt, the less humble or open-minded they are. Only that doubt which is truly humble, sincerely open-minded, should be labeled "agnostic." Only the man or woman who admits the possibility of being wrong is a humble agnostic.

Those taking the humble approach admit that the whole universe and all the creatures within it, both visible and invisible, may come from the eternal God and are manifestations of His infinite creative power. All of nature reveals something of the

Creator. And God is revealing Himself more and more to human inquiry, not always through prophetic visions or scriptures, but through the diligent research of modern scientists into observable phenomena and forces. The "golden age" of creation is being reached as God reveals Himself to human minds. As St. Paul said:

For what can be known about God is plain to them, because God has shown it to them. Ever since the creation of the world, His invisible nature, namely His eternal power and deity, has been clearly perceived in the things that have been made. (Romans 1:19–20 RSV.)

In the seventeenth century, the poet John Milton phrased it, "What if earth be but the shadow of heaven, and things therein each to other like more than on earth is thought?" Still another paraphrase in the language of modern science might be, "God is Creator of the universe of time and of men. Creation proceeds from idea to word to sensory data. The invisible Creator is the universal spirit, the causative idea which sustains and dwells in all He created and is still creating. The orderliness and lawfulness of nature and of the spirit reveal God to man."

God creates in many ways, not the least of which is through human endeavor. Like a pioneer in the wilderness, man must build the physical space in which he will live. As co-creator with God, he not only builds a material structure called "house" but imbues it with the spiritual dimension called "home." In one creative act, a log cabin, an Apollo spacecraft, or a cathedral contains both material and spiritual significance. The idea becomes a material fact; the idea, previously unseen, known only to God or to man, is now visible. Maybe the more we create, the more in some ways we are like God, especially if, like God, we create out of love.

For some more recent thoughts on "purpose," see our books *Evidence of Purpose* and *The God Who Would Be Known*.

IV The New World of Time

Most religions agree that God created the earth at the beginning of time. But now geologists, paleontologists, and astronomers assert that creation is a continuous process and might be still in its infancy. In other words, *this* is the beginning of time. If such is the case, theologians need to keep pace with scientists and revise the old notion that God created everything once and for all. As explained by N. M. Wildiers in *An Introduction to Teilhard de Chardin:*

Over a long period—up to a century ago, one might almost say—exploration of the universe invariably had the appearance of a venture into space . . . . Very gradually we have come to realize that the picture thus formed in our imagination had only a fleeting moment, a mere fraction of a long succession of changing circumstances. As has so often been said, the major discovery of modern science has really been the discovery of time—of time as a constituent of everything.[8]

We are different from our ancestors because of our modern concept of time.

But time is still a mystery. No one knows the nature of time or what it is. It could speed up or slow down without our being aware of it. Most theologians think that God is not subject to time, that He created or is creating it. Others speak of time as an aspect of the life of God. Certainly, God does not measure time by the rotations of one little planet called Earth in one little galaxy. Progress measured against time now appears to be accelerating; but remember the observation would be similar if instead time were slowing down.

In 1895, H. G. Wells wrote *The Time Machine.* This novel deals with a theme that explorers have dreamed about for centuries—the ability to travel backward and forward in time as though it were like travel over the face of the earth. Until only a thousand years ago, the only way to travel backward in time was imaginatively through history, which led one back only six thousand years. Then backward travel was extended by archaeology, paleontology, and geology to about six hundred million years. Recently, astronomers realized they could look backward into time through light signals traveling towards us from the past at only a hundred and eighty-six thousand miles per second. Quasars, some of which are estimated at twelve billion light-years distant, look as they did when the universe was young, because their light we see left them that long ago.

Time travel into the future is still not easy! Recently astronomers have developed theories about the birth and death of stars. Thus, by observing other stars earlier or later in their life cycles, we gain a mental picture of what our own sun may have looked like in its youth and will look like in its old age. The most frequent method of imaginary time travel is to observe a trend and then extrapolate that trend forward or backward into times unknown. This simple extrapolation is now producing weird and wonderful theories about the nature of the universe eighteen billion years in the past or in the future. This is a new kind of eschatology that would have astounded biblical writers.

Now that man has landed on the moon, it is natural to speculate about visiting or at least contacting civilizations far more advanced than ours on some planets of older stars. The Milky Way Galaxy consists of more than a hundred billion stars. There may be over a hundred billion other galaxies in the universe. If only one star in a thousand has planets and if only one in a thousand of these has a planet resembling the earth, the arithmetic still indicates the possibility of ten million billion other earths in the universe. In a statement made at the 1979 meeting of the American Association for the Advancement of Science, it was estimated that in our galaxy alone there may be millions of civilizations. Whatever we may say to that, most scientists agree that *Homo sapiens*, modern man, is not likely to be the end of evolution. If God's creative process is more advanced on some other planet, man may gain awesome and amazing revelations about his own future.

It would take us too far afield to describe here the various notions that men have formed through the ages regarding the universe as existing in time. It took us a very long time from the days when the earth was thought to be a huge disc floating upon the waters and covered over by a vaulted ceiling of stars, to reach the point where we could form a more accurate idea of its dimensions as well as its relationship to other planets in our galaxy and to all the other galaxies in the universe. Then, only by extensive and painstaking investigation and research was our modern conception of the universe existing in time finally achieved.

The old world views, however much they differed from one another, had certain things in common. Typical were their constricted dimensions, mechanistic structure, and static character.

Their constricted dimensions—the Ptolemaic picture of things—continued in vogue for more than a thousand years, up to the time of the Renaissance. In this generally accepted way of envisaging the cosmos, the earth was seen as a globe encompassed by hugh cystalline spheres. It was not until modern times that man became aware of the gigantic dimensions and enormous structure of the universe.

And, the old picture was explained by a mechanistic model of the universe. This meant that men saw the world as a combination of separate, heterogeneous elements "put together" extraneously with only a mechanical relationship to one another. A view of this sort made no proper allowance for the reciprocal cohesion of all entities. Just as a machine is made up of a number of previously prepared components, so the world was imagined to be composed of preconstituted and mutually independent entities that had been conjoined artificially. The earth, the vault of heaven, the plants, animals, and men were thus pictured as so many diverse "creatures" subsisting independently of each other and only making up a whole rather like, for example, pieces of furniture make up a living room.

In the modern-world picture there is a complete reversal of these conditions. Science has gradually made it more and more clear that all entities are continuously and intrinsically interconnected, so that we can now see the world as a mighty, organic whole in which every single thing is related to everything else. The world in which we live presents itself to us not as a machine, artificially contrived, but as an organism building itself up from within, an organism whose every part has appeared through a stage-by-stage process of growth.

Finally, in the old world picture, the universe was conceived of as a fundamentally changeless and static whole. Of course, men were not blind to the mutations and motions occurring in the world; but as they saw them, these changes were always on the surface of things and did not affect their essential nature. From its moment of origin, everything assumed a form and aspect that were definitive and unchanging, constant and unalterable. The machine was activated; it ran; but the machine itself never changed.

Along with the mechanistic view of the world, the old conception of it as static is now outmoded. Nowadays we see the universe as an enormous historical process, an evolution which has been going on for thousands of millions of years and is moving on into an incalculable future. The reason why the idea of evolution is of such great importance is that it points us to the fundamental and dynamic unity, or oneness of the world. Our world view, once static, has now become entirely dynamic.

Thus, there are three principle characteristics of the modern view: we live in a universe gigantic in its dimensions, building itself up organically as a cohesive whole, and impelled by an inner dynamic and energy toward its completion. The old idea of things has gone beyond recall, and now the world is revealed to us in a totally new guise. Only in the last generation have we begun to come to terms with the revolution that this has brought about in human consciousness.

Naturally, this new picture of the world did not spring up overnight, nor has it been the outcome of one particular branch of science. Small "pockets" of insight regarding the dynamic structure of things appeared here and there. Biology and its kindred life sciences have contributed even more than astronomy to forming this world view, for it is from the study of forms of life that the idea of evolution, of a process of progressive growth, is most clearly evident. From the life sciences the concept of evolution spread gradually to the other sciences, and in some degree influences our entire way of imagining the world.

The great pioneer in this field, of course, was Darwin, whose importance goes beyond his strictly scientific theories. It was Darwin who put the idea of evolution on the intellectual map once and for all. Since his time, it has been a constantly inspiring source of insight in all of human thought. The publication of *The Origin of Species* in 1859 will always remain one of the great turning points in the history of thought, cosmology, and eschatology.

In the wake of biology, the rest of the sciences has made the concept of evolution an integral element in their outlook and approach. In physics as well as in the sciences of the mind, we have come to see that we connot ultimately understand any phenomenon unless our investigation takes into account the way in which it has come to be what it is. We now know that even the atoms have their histories; that the stars have their birth, their prime, and their decay; that the languages have had their stages of development; that cultures come and go. The historical dimension of everything has become evident to us with unprecedented force, so that from now on, the categories of historicity extend over the totality of the universe.

All this suggests that the term "evolution" can be understood in two differing ways. By "evolution" we may mean the mutations that in the course of time have taken place in the various species of life (biological evolution). Or we may mean that the cosmos as a whole is subject to the law of evolution and that everything comes to be by a process of growth (cosmic evolution). We might say that biological evolution is only a segment of a far more comprehensive phenomenon: the evolution of the universe through time.

But, as advocates of the humble approach, we should pause to consider Sir Bernard Lovell's warning in 1978:

The complex processes leading to our present understanding of the universe have led to a modern view of the cosmos which we believe to be substantially correct, but it will be a remarkable and indeed unique feature of human thought if this really is the case.[9]

Nevertheless, at this moment in time, scientists, theologians, laity, all of us must utilize this current world view in our continuing participation in creation. Admitting we might be proven wrong, we launch boldly into new discoveries. And what new discoveries are to be made? Evolution may be not ending with man on earth but only beginning. To think that man on this planet is the end of evolution would be egotistical and anthropocentric indeed. It has always been difficult to imagine what comes next; but the multitude of discoveries in this century of things previously unseen points towards the likelihood of even more amazing discoveries hereafter.

How astounding it is that after ninety-nine percent of the evolution of the earth had taken place, a new creation, a new kind of evolution, could suddenly burst forth! The dramatic change was the appearance of man, a creature with free will, the first creature on earth to be allowed to participate in the creative process. Until then, evolution followed a course fixed by the laws of nature. But now suddenly the inconceivable has happened: self-evolution, intentional evolution has begun. The earth is filled with creations of a new kind—logic, love, mathematics, worship, purpose, inventions, and multitudes of other creations never seen on earth before. In reality after four billion years of earthly evolution, a new world was born, a world of mind.

According to the Jesuit paleontologist and mystic Teilhard de Chardin, in the long story of creation there came first the sphere of mineral evolution, the geosphere; then the sphere of living things, the biosphere; and lastly the sphere of the human mind, the noosphere. This unexpected new world of the human mind is so potent and so novel that no one knows what may happen next. Evolution is accelerating. The evolution of ideas is even faster than the evolution of new materials. There is no reason to think that this new world of mind and free will is the end of evolution. After the noosphere, there may be no "Omega Point" but a new sphere. What equally unexpected new world will emerge next? Will it be a new world of soul or spirit? Are we about to witness a new dawn?

Human ingenuity has caused revolutions in the geosphere and biosphere. Synthetic materials replace natural ones. In some parts of the world, we are rapidly stripping the land of resources in order to create standards of living never dreamed of by people a century ago. Luther Burbank began a revolution in the evolution of vegetables and flowers. Thousands of varieties which never existed before were invented by man to suit his needs and whims. Animals can now be bred with almost any desired trait. Some scientists think that by genetic engineering we can even produce animals with superior brains. More than ever men and women are participating in the creation of a new world.

Theologians need to grapple with some very hard questions. For example, the various methods of birth control can reduce the miseries caused by overpopulation, but we differ over the ethical issues involved in applying these new discoveries. Even more puzzling theological questions arise when we consider applying to human reproduction the new powers of selective breeding and genetic engineering. Does God want us to try to improve the human race not only through education and medicine but also by recombinant DNA experiments? Such rapid advances in science prove the urgent need for equal efforts toward spiritual progress and the need for new theologies. As St. Paul warned us:

Things beyond our seeing, things beyond our hearing, things beyond our imagining, all prepared by God for those who love him. These it is that God has revealed to us through the Spirit. For the Spirit explores everything, even the depths of God's own nature. (I Corinthians 2:9–10 NEB.)

Teilhard called for a new theology that would incorporate the modern scientific discoveries of the "immensity of space, which imbues our accustomed way of looking at things with a strain of Universalism" and the progressive "duration of time which . . . introduces . . . the idea of a possible unlimited Progress (Futurism)." Because of these two concepts, universalism and futurism, Teilhard believed we now possess a higher, more organic understanding of the cosmos which could serve as a basis for a new, unprecedented religion.

The twentieth century after Christ may very well represent new renaissance in human culture, a new embarcation into future cultures. Persons born in this century can hardly imagine the small amount of knowledge and the limited concept of the cosmos man had when the scriptures of all the five major religions were written. Do old scriptures need reinterpreting to accommodate an expanded notion of the universe?

More important for theology is the expanded concept of history. When all the scriptures of all major religions were written, the history of the universe was conceived as only a few thousand years. Now geologists and paleontologists who think in hundreds of millions of years read history in visible form sometimes more reliable than history books or scriptures. And cosmologists think in billions of years. Because light travels only a hundred and eighty-six thousand miles a second, we can see the sun not as it is now, but as it was some eight minutes ago. We see some stars as they were when Christ was born. We see some galaxies as they were six million years ago. Such a revolution in our conceptions of time and history is beginning to shape our theology.

What existed before this universe began? What will exist after the sun has grown cold? After minerals there emerged plants, and after plants animals, and after animals there emerged minds; and minds began to participate in the creative process. What comes next? Is there evidence that minds are developing into even more miraculous spirits and souls? These are not only questions of science but also of theology—a new type of theology not yet taught in the seminaries.

Consider the cold, inert world of minerals, the throbbing world of life, the curious, searching realm of the intellect. What next? This may be the most important question facing us at the end of the twentieth century. To answer it, scientists are daily engaged in new scientific experiments that will help us know more about the vast unseen. Theologians, too, answering Chardin's call for a new religion, must begin to explore the vast unseen dimensions of our evolving universe; they must plumb the very "depths of God's own nature."

V   Humble About What?

Humility is a misunderstood virtue. It can mean serving others, as when Jesus said, "Among you, whoever wants to be great must be your servant, and whoever wants to be first must be the willing slave of all." (Matthew 20:26–27 NEB.) But, humility toward men is not the subject of this book. Nor do we mean any belittling of the talents and blessings God has given us.

The word humility is used here to mean admission that God infinitely exceeds anything anyone has ever said of Him; and that He is infinitely beyond human comprehension and understanding. A prime purpose of this book is to help us become more humble and thereby reduce the stumbling blocks placed in our paths toward heaven by our own egos. If the word heaven means eternal peace and joy, then we can observe that some persons have more of it already than others. Have you observed that these are generally persons who have reduced their egos, those who desire to give rather than to get? The Holy Spirit seems to enter when invited and to dwell with those who try to surrender to Him their hearts and minds. "Behold, I stand at the door and knock; if any man hear my voice and open the door, I will come in to him and sup with him, and he with me." (Revelation 3:20, KJV.) As men grow older and wiser, they often grow in humility.

The humble approach has much in common with but is not the same as natural theology, process theology, or empirical theology, whose horizons are all too narrow. They often attempt to give a comprehensive or systematic picture of God in keeping with human observations. But the humble approach teaches that man can discover and comprehend only a few of the infinite aspects of God's nature, never enough to form a comprehensive theology. The humble approach may be a science still in its infancy, but it seeks to develop a way of knowing God appropriate to His greatness and our littleness. The humble approach is a search which looks forward, not backward, and which expects to grow and learn from its mistakes.

All of nature reveals something of the Creator. That golden age of creation is reached as the Creator reveals Himself more and more to the minds of men. Men cannot learn all about God, the Creator, by studying nature because nature is only a contingent and partial manifestation of God. Hence Natural Theology which seeks to learn about God through nature is limited. Recently a new concept of theology has been born which is called Theology through Science. This denotes the way in which natural scientists are meditating about the Creator on the ground of their osbervations of the astronomic and subatomic domains, but also on the ground of investigations into living organisms and their evolution, and such invisible realities as the human mind.

Experimental theology can reveal only a very little about God. It begins with a few simple forms of inquiry, subject to little disagreement, and proceeds to probe more deeply in thousands of other ways. Spiritual realities are not quantifiable, of course, but there may be aspects of spiritual life which can be demonstrated experimentally one by one even though there be hundreds of failures for each success. This approach is similar to that of experimental medicine.

As with experimental theology, the humble approach implies that there is a growing body of knowledge and an evolving theology not limited to any one nation or cultural area. The truly humble should be so open-minded that they welcome religious views from any place in the universe that is peopled with intelligent life. Seekers following the humble approach are never so zenophobic that they reject ideas from other nations, religions, or eras. Because the humble approach to theology is ongoing and constantly evolving, it may never become obsolete.

When learning about God, a world-wide approach is much too small. Even a universe-wide approach is much too small. The "picture" ninety-nine percent of people have of God is small. Have you heard anyone say, "God is a part of life"? Would it not be wiser to say of humanity that it is only an infinitesimal speck of all that has its being in and through God? Our own ego can make us think that we are the center rather than merely one tiny temporal outward manifestation of a vast universe of being which subsists in the eternal and infinite reality which is God. Surely life is a part of God, not vice versa. Have you heard the words, "the realm of the Spirit"? Is there any other realm? Humanity on this little earth may be an aspect of all that is upheld by the Spirit, but the Spirit is not an "aspect" of humanity. To say that God is a "part" or an-"aspect" of life is as blind as for a man, standing on a shore looking at a wave, to say, "The ocean is an aspect of that wave."

The ocean-wave analogy is helpful in approaching God with more humility. A wave is part of the ocean, having no existence apart from the larger body of water. The wave is temporary, whereas oceans are relatively permanent. Each wave is different from each other wave. In a sense, the wave is created by the ocean and is a child of the ocean. When it dies, it returns to and continues to be a part of the surging oceans creating ever new breakers on the beach. It is this keen sense of proportion and relationship that those advocating the humble approach seek to encourage.

The humble always respect the hierarchy of being evident in creation. Consider a tree. The tree is alive, having been created just as we were. Like us, it consists of billions of cells and atoms. It ages, dies, and hopefully produces seeds after its kind. But the tree cannot describe us. It has no ability to comprehend us or the complex culture of which it is a mere part. It may be a beautiful addition to our garden. We may nourish and care for it. But the difficulty the tree would have in describing us is perhaps similar to our difficulty in writing an account of God. If any person were to say he knew all about God, it would be like a tree claiming to know all about its gardener. Those following the humble approach maintain a belief in the great ladder of being.

Similar to the chain of being, with its unending variations and surprises, is the mosaic of human opinion about God. The sheer variety of opinions concerning the nature of God should give us pause. We should admit humbly that other theologies contain valuable insights into God that our own may lack. In many ways, men attempting to describe God encounter the same difficulties as the proverbial blindmen describing an elephant. No one sees the entire picture, yet each feels and understands a part necessary for the full description.

To have all men believe alike would be a great tragedy. Progress would cease. The spiritual struggle would be over. Life would hardly be life. The more we know, the more we know we do not know. This is what gives life spice. In fact, in order to grow, we must daily become more humble and honest in admitting the paucity of our knowledge. This humble admission of ignorance is what produces progress, what keeps man searching, what makes life as we know it exciting and challenging.

Gaining knowledge is like working a quarry. As we chip out bits of information, the mining face gets larger and larger. The more knowledge we gain the more we can see the extent of the unknown. As we grow in knowledge, we grow in humility. This may be just as true in studying the soul as in the investigations pursued by natural sciences.

A man or woman pursuing the humble way to God could adopt a credo similar to the following:

God is billions of stars in the Milky Way and He is much more.

God is billions of billions of stars in other galaxies and He is much more.

Time and space and energy are all part of God, and He is much more.

The awesome mysteries of magnetism, gravity, light, knowledge, imagination, memory, love, faith, gratitude, and joy are all part of God and He is much more.

God is five billion people on Earth and He is much more.

God is untold billions of beings on planets of millions of other stars and He is much more.

God is all the things seen and also the vastly greater abundance of things unseen by man.

Men who dwell in three dimensions can comprehend only a little part of God's multitude of dimensions.

God is the only reality—all else is fleeting shadow and imagination from our very limited five senses acting on our tiny brains.

God is beginning to create His universe and allows each of His children to participate in some small ways in this creative evolution.

God is the infinitely large and also the infinitely small—He is each of our inmost thoughts, each of our trillions of bodily cells, and each of the billions of wave patterns which are each cell.

God is all of you and you are a little part of Him.

By constant reminders such as this, we might avoid the most common pitfall of theology: the attempt by man to put limits on God.

Throughout all history, the gods created by men's minds have been too small. All concepts knowable to man have the specific characteristics of the human thought which formulates them. How often have we heard someone deny the existence of God because good people suffer. It is as if that unbeliever attributes to God "man's inhumanity to man" and assumes that "God's inhumanity to man" disqualifies Him from being God! Does not God work in mysterious ways? Cannot what appears to us as "God's inhumanity" or injustice be part of a more perfect picture that we do not see? God's love is not as small as ours. It is infinite. In fact, if we can never fully understand even another human being, how will we ever understand God?

It is a great mystery that evolution has provided us with the mental ability to think about our Creator. As far as we know, only humans among all of God's creatures on earth have the ability to think about God. But all concepts we can form are limited by our five senses and by the smallness of our minds. Therefore we can never totally know God. We really only know our attempts to know Him; we devise theologies hoping they in some way adequately represent Him. And yet they are always inadequate. If God were small enough to fit our human reason, He would not be God at all but only another human. In some ways, we are like radio receivers that can receive music, voices, and wonderful sounds from hundreds of sources but are hopelessly blind to the sunsets and flowers.

To each of us has been given free will and a mind which is itself a creative power. We can create only in a very limited way. However, because our ability to create and to understand appears vastly greater than that of any other kind of earthly creature, we are thought of as made in the image of God.

Many religions hold that knowledge about God comes not so much from human reasoning as from God choosing to reveal Himself to us. The Church of Christ of Latter Day Saints, for instance, is based on revelation, translated in 1828, from books of gold. The Prophet Muhammad (570–632 A.D.) received revelations which were memorized and later written down by his disciples.

Christians believe God came into the world two thousand years ago and revealed Himself in the life of Jesus of Nazareth. But none of His words and no accounts of his life were evidently written down until a generation after his ascension into heaven.

Men could write down only what they understood. Communication reduced the revelations to the mental development of the messengers. If God had revealed the General Theory of Relativity two thousand years ago, no one would have been able to write it down. Maybe God is ready to reveal Himself more and more. Maybe by evolution each generation is able to comprehend a little more. Ninety-nine percent of God's universe has become known only in the latest half-century. Nuclear physicists and geneticists and astronomers are earnestly seeking with open minds and with humility. Can you say the same for all churches and all theologians?

There are clear scriptural bases for advocating the need for an inquiring and open mind. According to St. Luke, Jesus said, "Ask, and it shall be given you; seek, and you shall find; knock, and it shall be opened unto you. For everyone who asked, receives, and he who seeks, finds; and to him who knocks it shall be opened." Maybe God reveals Himself where He finds an inquiring mind—an open mind. In the Acts of the Apostles (17: 24–28, NAS) St. Paul said:

The God who made the world and everything in it . . . made from one every nation of men . . . that they should seek God, in the hope that they might feel after Him and find Him. Yet He is not far from each of us, for "In Him we live and move and have our being"; as even some of your poets have said, "For we are indeed His offspring."

Christ came to reveal God to men. But because of the limitations of human minds and human language, maybe less than one-hundredth part has been handed down to us. It is easy for us to realize how ignorant and primitive were the Jews of two thousand years ago and the Hindus of three thousand years ago. We should be humble enough to admit that if they had only perhaps one-tenth of one percent of all knowledge, we may have only one percent, even though the little glimpses we do have are indeed awesome.

One following the humble approach thinks it possible that God may want to reveal Himself further than He has done to date in any major or minor religion. He may be ever ready to give us new revelation if we will but open our minds to seek and inquire, but first we must rid ourselves of that rigidity and intellectual arrogance that tells us we have all the answers already. Like natural scientists who already assume the humble approach in their studies, maybe we should recognize that the law of creation is a law of accelerating change. Human language has always been too inadequate and restricted to utter all truths once and for all. The human mind has never been ready to receive all knowledge.

Time, space, and energy are the limits of our lives as they are the limits of our knowledge. God, of course, is not bound in these ways. He is the Creator of the awesome vastness of His cosmos. He knows each person's most fleeting thought just as

He knows the power of a quasar and the intricate complexity of a DNA molecule. His most marvelous and mysterious creation on earth is the human brain with its indwelling mind. With the use of our minds, we can participate in some small ways in the creation of matter and even life itself. It should be clear to us that even though we are seriously hampered by our human weaknesses, we are meant to share with God his readiness to reveal Himself to us. We have a duty of humility, the duty to be open-minded.

In humility, we should try to be continually creating our own personalities and souls by learning more about the Creator. St. Paul says in I Corinthians (2: 9–12, NEB):

But in the words of scripture, "Things beyond our seeing, things beyond our hearing, things beyond our imagining, all prepared by God for those who love Him," these it is that God has revealed to us through the Spirit. For the Spirit explores everything, even the depths of God's own nature. Among men, who knows what a man is but the man's own spirit within him? In the same way only the Spirit of God knows what God is. This is the Spirit that we have received from God, and not the spirit of the world, so that we may know all that God of His own grace has given us.

The human body is even more awe-inspiring than the stars of the universe. We recall again that astronomers now think that the Milky Way galaxy contains over 100 billion stars and that the universe contains over 100 billion galaxies. The number of stars in the universe is truly astronomic: but the number of atoms in our body is even greater, about 5 million times as great.

An average cell may be made of 50 thousand billion atoms. The average human may have 100 thousand billion cells. So the atoms in our body may total 5,000,000,000,000, 000,000,000,000. We each have about thirty billion cells in our brain alone. Some cells are spheres, some spirals and rectangles, others have tails. Some nerve cells are star-shaped with tiny bodies trailing wispy arms. Consider the astronomer Carl Sagan's description of the human brain:

If each human brain had only one synapse—corresponding to a monumental stupidity—we would be capable of only two mental states. If we had two synapses, then $2^2 = 4$ states; three synapses, then $2^3 = 8$ states, and, in general, for N synapses, 2^N states. But the human brain is characterized by some 10^{13} synapses. Thus the number of different states of a human brain is 2 raised to this power, i.e. multiplied by itself ten trillion times. This is an unimaginably large number . . . . It is because of this immense number of functionally different configurations of the human brain that no two humans, even identical twins raised together, can ever be really very much alike. These enormous numbers may also explain something of the unpredictability of human behavior. The answer must be that all possible brain states are by no means occupied; there must be an enormous number of mental configurations that have never been entered or even glimpsed by any human being in the history of mankind. From this perspective, each human being is truly rare and different and the sanctity of individual human lives is a plausible ethical consequence.[10]

Thus, if as humans we continually surprise ourselves, should we not be prepared to be surprised by God?

So what is it we are to be humble about? Our experience of God, our experience of ourselves? Perhaps the experience of God that we overlook most easily is the experience of Him through science. Being humbled before science is a good first step toward the humility we should have before God. As Vannevar Bush puts it:

Science here does things. It renders us humble and it paints a universe in which the mysteries become highlighted, in which constraints on imagination and speculation have been removed, and which becomes ever more awe-inspiring as we gaze . . . on the essential and central core of faith. Science will be the silence of humility not the silence of disdain.[11]

Just as we need the humility to live with conflicting scientific theories, we also need the humility to accept conflicting religious attitudes. Friedrich Schleiermacher wrote in 1799, "Nothing is more unchristian than to seek uniformity in religion." He argued that instead of being segregated behind ecclesiastical walls, differences of opinion should be allowed to work upon, enrich, and define each other. He claimed that lack of appreciation for other religions has no basis in religion itself. The more adequately one learns about the infinite, the more humility and openness one will have for the unlimited range of its manifestations.

Maybe the Creator's plan (not only for the physical universe but also for the metaphysical) is ever-increasing revelation, growth, change, and variety.

VI   Benefits from Humility

Humility is the gateway to understanding. As thanksgiving opens the door to spiritual growth, so does humility open the door to progress in knowledge and also to progress in theology. Humility is the beginning of progress. Humility leads to open-mindedness. It is difficult for a person to learn anything more if he is sure he knows it all already. When we realize how little we know, we can begin to seek and to learn. Unless we recognize our ignorance, why should we investigate?

For each of us to grow in spirituality, we should be humble in worshiping God. If we free ourselves of self-will and surrender to God's will, we can become channels for God's love and wisdom to flow to others. Theologians have long proclaimed this; but sometimes, when formulating creeds or describing God, they seem to deny it. Some sound as if they think they know it all. This can lead to statements that make God appear limited and anthropomorphic, rather than truly infinite and majestic in His divine power.

For any theologian or minister to claim he knows all about God, when he really knows less than a minute amount of what may be known, makes him appear arrogant and obscurantist to educated listeners or readers. And worse, it acts as a closing of the doors so that the fullness of God's light cannot flow freely through him to others. A closed mind does not produce progress. An atheist who is sure there is no God is really a pitiful person because he is too egotistical to admit his limitations and insignificance. Because his small concept of God may not seem to fit with facts observed by scientists, he denies that God exists rather than admit that his own concept is too small. How small is a man who says, "If God does not conform to my thoughts, then He does not exist"? Before he can receive the Holy Spirit he must surrender his pride and "become as a little child."

Some people have said that religion causes wars, that the diversity of religious beliefs will always be divisive. But the fact is that such wars were brought about for a number of reasons, not the least of which was egotistical and unwarranted claims to proprietary rights over the knowledge of God. History reveals that great havoc and suffering were caused not by religion, but by men who thought their concept of God was the only one worthy of belief. "He that is of a proud heart stirreth up strife," it says in the book of Proverbs (28:25, KJV). Pride was the first of the 7 deadly sins enumerated by Saint Gregory. The verbal and physical violence in religion could be reduced by believers who claim nothing more than a minimal knowledge of the infinity of God.

Ralph Waldo Trine has said, "Let us not be among the number so dwarfed, so limited, so bigoted as to think that the infinite God has revealed himself to one little handful of his children, in one little quarter of the globe, and at one particular period of time."[12] In fact, at the heart of true religion is the willingness to see truths in other religions. The Persian scriptures claim, "Whatever road I take joins the highway that leads to Thee . . . . Broad is the carpet God has spread, and beautiful the colors he has given it." A Buddhist believes that, "The pure man respects every form of faith. My doctrine makes no difference between high and low, rich and poor; like the sky, it has room for all, and like the water, it washes all alike." A Chinese sage has said, "The broad-minded see the truth in different religions; the narrow-minded see only the differences." And a Hindu holy man has written that only the narrow-minded ask, " 'Is this man a stranger, or is he of our tribe?' But to those in whom love dwells, the whole world is but one family." Lastly, the Christian witnesses to this with the following words: "Are we not all children of one father? God has made of one blood all nations to dwell on the face of the earth."

Differing concepts of God have developed in different cultures. No one should say that God can be reached by only one path. Such exclusiveness lacks humility because it presumes that we can and do comprehend God. The humble person is ready to admit and welcome the various manifestations of God.

Jesus quoted Isaiah thus: "But in vain they do worship me; teaching for doctrine the commandments of men." (Matthew 15:9, KJV.) Schism in religions is caused by intolerance; and intolerance is a form of egotism. However, tolerance is not the same as the humble approach. We should seek to benefit from the inspiring highlights of other denominations and religions, not just to tolerate them. We should try our very best to give the beauties of our religion to others, because sharing our most prized possessions is the highest form of "Love thy neighbor." Let us not water down the diverse religions into a know-nothing soup; but rather let us study enthusiastically the glorious highlights of each. An old Chinese precept is, "The good man does not grieve that other people do not recognize his merits. His only anxiety is lest he should fail to recognize theirs." It is a mistake for people of different religions to try to agree with each other. The result is not the best of each but rather the watered-down, least-common denominator. What is more fruitful is a spirit of humility in which we recognize that no one will ever comprehend all that God is. Therefore, let us permit and encourage each prophet to proclaim the best as it is revealed to him. There is no conflict unless the restrictive idea of exclusiveness enters in. We can hold our ideas of the Gospel with the utmost enthusiasm, while humbly admitting that we know ever so little of the whole and that there is plenty of room for those who think they have seen God in a different way. The evil arises only if one prophet forbids his audience to listen to any other prophet. The conceit and self-centeredness of such restriction—the false pride of saying that God *can* be only what we have learned Him to be—should be obvious.

The human ego has been the curse of religious denominations for thousands of years. In every major religion wars have been fought about differences of creeds. Nations or tribes have exterminated others because they worshiped different gods or the same god as taught by different prophets. This is human ego run wild. Let us humbly admit how very small is the measure of men's minds. This realization helps to prevent religious conflicts, and obviates attacks by atheists against religion. Moreover, humility of this kind opens more minds to the idea that science supports and illuminates religion.

If a person concentrates only on his progress or enlightenment, he is essentially self-centered; for he should also help others, and share what he has learned with others by speaking, writing, broadcasting, and disseminating his ideas. Such new evangelists should be welcomed and respected for their generosity.

Influenced by humanism, some clergy in Christian churches unknowingly encourage religious strife by working for political or social goals rather than purely spiritual ones. They strive too much to create an earthly, material, man-made kingdom in contradiction to the spiritual, internal kingdom that Jesus preached. In Third World nations, for example, churches sometimes advocate armed force by governments, or even violence by terrorists for the sake of social justice. They neglect to teach their members that enduring improvement must begin with self-improvement and voluntary improvements in our own hearts. Compulsory brotherhood is a contradiction in terms.

Christians have sometimes derived their compulsory, intolerant attitudes from a statement such as this one in St. John. Jesus said, ". . . no one comes to the Father except by me. If you knew me you would know my Father too." (John 14: 6–7, NEB.) But is this an intolerant and exclusive claim? At first glance, it is. But perhaps we can see what Jesus meant by recalling a similar statement in Matthew, "No one knows the Son but the Father, and no one knows the Father but the Son and those to whom the Son may reveal him." (Matthew 11:27, NEB.) Jesus prefaced this comment with the words, "I thank thee, Father, Lord of heaven and earth, for hiding these things from the learned and wise, and revealing them to the simple." What he seems to be saying is that there are frontiers to human knowledge of God, but these frontiers are open to the lowly and simple of any age or culture. Proud and overbearing egotism prevents knowledge of the Father, and is clearly opposite to the teachings of Jesus.

Christians believe that God came into the world in human form to reveal Himself. But human intelligence is so limited that we could understand only a little. Missionaries have difficulty revealing all the gospel to primitive natives, so that they reveal just what the natives can comprehend. Jesus, Himself, we are told, taught the multitudes only as they were able to hear and understand. He found human language and philosophy two thousand years ago so inadequate that He used symbolism, metaphor, living examples, parables, and so on. Maybe the reason why He wrote nothing is that writing was and is inadequate to convey truth without warping and constricting it. For example, suppose a man so loved the forest that he transformed himself into a tree and then tried to use the limited language of trees to teach them about man. The trees could form only small concepts of man and their contact with humans in various ways could produce merely a variety of small concepts. Is not the superiority of man over trees infinitely less than the superiority of God over man?

Some people do appear to come closer to God when they pray in Jesus' name, possibly because they have progressed more in that upward path of the humility and meekness of Jesus than others. But on the other hand, some appear more spiritual when they pray as disciples of Buddha or Mohammed or Abraham. Theodore Parker taught that the doctrinal formulations of Christianity have changed and will change from age to age and what is sometimes called heresy at one time is accepted as orthodox and infallible in another age. Old forms, Parker says, give way to new, and each new form will capture some of the truth but not the whole. "Transient things form a great part of what is taught as religion. An undue place has often been assigned to forms and doctrines, while too little stress has been laid on the divine life of the soul, love to God and love to man." [13]

Religious advance springs from deep humility. Therefore, it comes most easily to a person who is aware of his ignorance and his need to learn. It is more difficult for a man highly educated in theology to comprehend how infinitely minute his knowledge is compared to God's knowledge. It is significant that Jesus was not trained as a scribe or pharisee. St. Luke recorded:

Jesus looked at him and said "How hard it is for those who have riches [or education or intellect] to make their way into the kingdom of God! Yes it is easier for a camel to pass through the eye of a needle than for a rich man [or an intellectual] to enter the kingdom of God." (Luke 18: 24 and 25, JB.)

Intellect and wealth both seduce men into self-reliance. Of course some theologians do know ten times as much about religion as the average layman or minister. But if they claim to have more than one percent of knowledge of God, they are ridiculous in their pride. St. Paul said:

Make no mistake about this: if there is anyone among you who fancies himself wise, I mean by the standards of this passing age—he must become a fool to gain true wisdom. For the wisdom of the world is folly in God's sight. Scripture says "He traps the wise in their own cunning"; and again, "The Lord knows that the arguments of the wise are futile"; so never make mere men a cause for pride. (I Corinthians 3:18–21 NEB.)

The human ego makes people try to solve problems by human effort alone without turning to seek assistance in God's wisdom. Rigid creed is a form of pride, for it means we think we understand all about God. Both the Old and New Testaments are critical of the proud. "Everyone that is proud in heart is an abomination to the Lord," it says in Proverbs (Proverbs 16:5, KJV), and St. James warns, "God resisteth the proud, but giveth grace to the humble." (James 4:6 KJV.) "A man's pride brings him humiliation; he who humbles himself will win honor." (Proverbs 9:23, KJV.) "Pride goeth before destruction and a haughty spirit before a fall." (Proverbs 16:18 KJV.) "By humility and the fear of the Lord are riches, and honour and life." (Proverbs 22:4, KJV.) "Before honour is humility." (Proverbs 15:33, KJV.) "With the lowly is wisdom." (Proverbs 11:2, KJV.) "Humble yourself in the sight of the Lord and he shall lift you up." (James 4:10 KJV.)

Sometimes large councils of churches make exclusive dogmatic pronouncements, even in fields where they are poorly informed, such as economics and politics. This can create division, hatred, or strife. Perhaps results would be better if they expressed love for all, welcomed diversity, and avoided the sin of self-righteousness.

Becoming "unselfed" opens the door to communication with God. He who relies on his own wisdom or beauty or skill or money shuts God out. But he who is humble and grateful for such God-given blessings opens the door to heaven on earth here and now. For each of us to grow in spirituality, we should be humble in worshiping God. We should free ourselves of self-will and surrender to God's will. If we get rid of ego-centeredness, we can become clear channels for God's love and wisdom to flow through us, just as sunlight pours through an open window. In the language of electronics, man is a receiver and transponder; but when his pride causes him to conceive of himself as a transmitter instead, he naturally cuts himself off from God who is his source of supply for both love and wisdom.

We should be as humble as the scientist investigating nature, limiting himself to those parts of nature which can be observed. In fact, we should take as our models those careful scientists, conscientiously differentiating between the few aspects of reality that they can observe and those vast uncharted areas for which they have not yet devised technologies and methodologies for research. We should imitate those scientists who are not deluded by intellectual pride and doubt and who do not deny the metaphysical aspect of life. Scientists following the humble approach to science as well as to God would never assert that what they cannot comprehend is for that very reason nonexistent. The scientists, however, who have surmounted this barrier of human egotism are beginning to investigate and learn about unseen realms, including the spiritual.

As we have seen earlier, perhaps both scientists and theologians should pool their humility and explore together the distant corners of the universe. Why should either group, in its arrogance, design a small pattern of thought and insist that God and His creation fit into it? For example, in the distant future it may be regarded as incredible that in those dark ages called the twentieth century some medical doctors were so limited as to deny that any benefits could come from the prayers of Christian Science practitioners; and that some Christian Scientists were so limited as to deny that God might choose to work through the hands and minds of medical doctors. Such narrowness is the opposite of the humble approach. More and more healing and health seems to come from the inside as well as the outside; but why not use spiritual and material remedies both together.

When the disciples of any cause, sacred or secular, stop seeking and claim they have the answer, the movement ossifies. It appears that God's creative method is movement, change, continuing search, ongoing inquiry. Those who seek are rewarded. Those who are sure they already have the answers gradually become obsolete. Perhaps "built-in" obsolescense is God's plan for keeping the world of ideas forever young, fresh, and invigorating. The self-confident proud grow old and die, and with them die their ideas.

Instead of burning the heretic at the stake, we may benefit more if we listen to him and carefully observe whether his ideas bear good fruit or not. If his message is not holy, it will fade away when subjected to the free competition of ideas. The open-minded approach is to look for God in a multitude of ways, in the kind of empirical questioning pursued by natural scientists and those theologians who recognize that some of tomorrow's spiritual heroes may be among those considered today as heretics.

Progress occurs most often in a condition of free competition. God gave us free will. He might have chosen to create us as angels with all the knowledge we would ever need or want.

But in His wisdom He chose to give us free will with which to desire and long for, to try, to fail, to succeed, and finally to progress. It is contrary to this divine purpose for any church or national government to dictate that we must all fit into one mold, to take away free will, to eliminate the free competition of ideas. Any large government enforcing regimentation retards progress and change, and reduces souls to machines.

New discoveries come mainly from free minds. The worship of God is not weakened by keeping an open mind any more than cosmology was weakened when Ptolemaic theories were replaced by those of Copernicus. Maybe one of the attributes of God is change. Did He decree the survival of the fittest? Maybe God intends us in some way to use the new power He has put into our hands in relation to selective breeding and recombinant DNA, to improve the human race. Why else does He give us this awesome new knowledge and the free will to use it?

By learning humility we find that the purpose of life on earth may be far different from what anyone now supposes. Diligently each child of God should seek to find out and obey God's purposes. If we encourage free and friendly competition of ideas, the truth may more easily emerge. It may be God's plan that any organization that tries to maintain its "truth" as the status quo is headed for extinction. No matter how well protected, every willow tree grows old and dies. But if, before it dies, a little limb is snipped and put in good moist soil, it will take root, flourish, and grow into another tree. This principle seems to apply to plants, animals, people, nations, churches, and even to ideas. From a little sprig of an idea, the world's great bodies of thought grew year by year until they were fully mature. For example, despite harsh persecution the early Christian church grew rapidly like a young willow tree for five centuries. Likewise for five centuries, the ideas of Mohammed swept the world like wildfire. Which religions and denominations are growing most rapidly today? Many with the most rapid growth are those which are still young, still exploring, still taking risks, still experimenting with the meaning of life.

Secular ideas and institutions also show this tendency toward the life cycle. Communism, for example, captured a quarter of the world's people in one century for many reasons, a primary one being its youthful promise to sweep away the stifling patterns of older societies. But in many nations, socialism and communism became so restrictive and coercive that a growing number of dissidents refused to tolerate it. People criticize or leave their countries for other places that permit and encourage free expression of thought.

Youthfulness is an important principle of life whether considered from a biological or spiritual viewpoint. The young at heart or the young in mind are humble and open-minded and welcome new ideas. Jesus' admonition to become like little children may have been His way of telling us to experiment, explore new ideas, test things, admit our lack of knowledge and know-how, and to be humble. Yes, change and progress require a youthful attitude. Ideas and institutions which remain rigid for generations tend to wither.

Christians think God appeared in Jesus of Nazareth two thousand years ago for our salvation and education. But we should not take it to mean that education and progress stopped there, that Jesus was the end of change, the end of time. Is such notion compatible with God's law of the universe? To say that God cannot reveal Himself again in a decisive way because He did it once years ago seems sacriligious. We should be gentle, kind, and sympathetic toward God's new prophets even though they bring strange new ideas. We should not forbid religious expression, however misguided we think it to be. No useful purpose was served when the Inquisition forced Galileo to recant in 1633. Jesus, too, was considered unorthodox by the learned religious establishment of his day. Those with the humble approach invite the new evangelists to share what they think they have learned with others by speaking, writing, broadcasting, and publishing their ideas.

Just as each tree or child or soul can grow in the correct nourishing environment, so too, could the world-soul of humanity grow and flourish if we had global toleration and inquiry. Religious leaders in every nation could increase spiritual understanding just as fast as scientists increase our understanding of the physical world. A lot, of course, depends upon our willingness to love those who are strange and different from ourselves even though they cling to their strange and new ideas. As Ella Wheeler Wilcox put it:

A thousand creeds have come and gone
But what is that to you or me:
Creeds are but branches of a tree
The root of Love lives on and on.

Humbly to admit that we know only a very little of God's truth does not make us agnostic. If a medical doctor can admit with an open mind that he does not understand all diseases, symptoms, and cures, surely we can be equally humble and honest, not agnostic, by admitting we each have more to learn about God. We ought not to brand every skeptic as a godless materialist, any more than we would have skeptics accuse every believer of being a dogmatic roadblock to progress. There is room for many branches on the tree. The life-sap of love lives on and on.

VII Creation Through Change

Creation is just beginning. We are just starting to understand that God has given us talents so we can participate in His creative process. The old ways of structuring and ordering institutional religions may not apply to the future. Often they are too rigid, too traditional. New, freer, more imaginative and adaptable creeds will have to be devised in order that man's God-given mind and imagination can help to build the kingdom of heaven.

If, as Teilhard said, the universe is "a huge psychic phenomenon," an upward evolution toward increasing consciousness made possible by ever more complex structures, we must somehow encourage our more gifted individuals to dedicate their time and talents to this ongoing work of creation. We will need the best minds and hearts. In the whole sweep of evolution we see movement from the simple toward ever richer complexity and variety. So, too, human creators must represent the wide variety of human thought and invention. Men should try to produce creations that truly reflect the talents God gives us. Just because man *seems* to be the highest species of creation, the end of cosmogenesis, we should not be deluded into thinking we are the lords of creation. We are the servants of creation. We are a new beginning, the first creatures on earth allowed to participate consciously in the evolving creative process.

Today the world urgently needs new breakthroughs for man's basic understanding of God. Theologians need to be humble and open-minded. It is good for churches to have creeds, doctrines, dogmas, liturgy, and the hierarchies of layman and clergy. This helps the church to exist as an organization of people whose ideas are compatible. This gives continuity to the development of the church; and it helps outsiders to see what each church denomination represents. However, because of a lack of humility, there is today and always has been throughout the history of all churches a tendency for dogma or hierarchy to stifle progress. If the members and clergy were more humble, they could use dogma in a more open and inquiring way as a beginning point for continual revision and improvement. If a family (or a corporation or church) prepares a good budget but never changes it for a century, such a budget will surely become obsolete and gradually change from being a help to being a millstone hung round the family's neck. Budgets are helpful only if they are continually revised and improved. Like budgets, church structures are more helpful if they are continually revised and improved. John Calvin taught that "to be Reformed is to be always reforming."

It is interesting that throughout history religion has developed and progressed most often by the work of those who were first regarded as heretics. The pharisees were learned and holy men, but most of them seemed to have regarded Jesus as a heretic. Other heretics were Buddha, Paul, Zoroaster, Mohammed, Wycliffe, Hus, Luther, Calvin, Wesley, Fox, Smith, Emerson, Bahaullah, and Eddy. Christians believe God chose to enter the world in human form but not as a traditionalist urging merely restudy of Abraham and Moses but rather as a progressive with a new revelation. Rarely does a conservative become a hero of history. Rather, it is usually a progressive, far-reaching thinker, one who breaks out of a traditional mold. In other words one who, according to the accepted customs of his time, would be branded a heretic. The opening chapters in the life of most good heretics are usually about humility; the desire to learn more, and eventually the growing awareness that something innovative should be tried.

The earth needs a new Columbus, a new Galileo, a new Copernicus, a genius who can enlarge the global vision of mankind to reveal how tiny and temporary we are in comparison to the infinity and eternity of God. Who can make us humble enough to comprehend that mankind may not be the end of the creative process, nor the earth the center of the universe?

Dr. Robert Hilliard has said that if the growth of knowledge continues at the present rate, then when a baby born today graduates from a university, the quantity of knowledge on earth will be four times as great. By the time such a child is fifty years old the amount of knowledge will be thirty-two times as great; and ninety-seven percent of everything known to man will have been learned since the day the child was born. John Naisbitt says the quantity of information is doubling every 30 months. At such rate the quantity may be 1000 times as great in only 25 years.

This is the blossoming time in the creation of man. Evolution is accelerating. Progress is accelerating. One of God's great blessings to man is change, and the present acceleration of change in the world is an overflowing of this belssing. Those who love God should devote as much manpower and as many resources to research in spiritual subjects as corporations and governments devote to research about material things. Seminaries need to do as much research as universities.

Only about one child in a million is born with a mind that is superhuman in one or more ways. Why does God's process of evolution produce these rare geniuses on earth? Is it His plan that they should help all people to progress? The one in a million who contributes a new idea to humanity can be a blessing to millions, so that God's creation can continue to progress. God gave each oyster the ability to have a million children of which only one, on the average, reaches maturity. Is it the purpose of the million to ensure that one will preserve the race? In addition to the geniuses given more-than-human minds, God also creates saints and prophets with more-than-human souls. A prophet is a pioneer in the uncharted regions of the spirit. No two persons are equal or identical in body or mind, and it is probable that no two persons are equal in soul.

However much we may yearn for equality, it does not seem to be part of God's plan.

Sören Kierkegaard taught that the human race advances on the backs of those rare geniuses who venture into realms that most of us are afraid of. Arend van Leewin has said, "Ninety-nine percent of people, irrespective of race, play a passive as opposed to a creative role; and even the creative section are passive with regard to ninety-nine percent of their civilization."[14] And Huston Smith, the masterful chronicler of world religious thought and practice, wrote:

The average man is no more capable of forming his imagination in ways that resolve his feelings nobly than he is capable of being his own scientist. Both tasks require genius. Geniuses in the art of shaping man's imaginings are artists, philosophers, prophets, and seers. Over time their creations coalesce and distill into cultures. As the religious forms of traditional Judaism and Christianity are losing their powers to inform the contemporary mind, the West desperately needs religious geniuses who can create new imaginal forms, convincing to the contemporary mind, which consummate man's needs for home, vocation, and transcendence.[15]

Any tendency to stifle religious genius by adopting excessively detailed religious laws or unchanging doctrines, liturgies, or structures is the result of a lack of humility. Histories of all religions show that traditionalists are eventually hopelessly out of touch with life, and that their position usually passes away. Fixed dogma and bureaucracy are similar. Both squelch progress. Both may ultimately impede the path of each human soul into heaven. If a purpose of man on earth everywhere is to be a co-creator with God, it is presumptuous or egotistical for any one group to say that there is only one kind of believer whom God can use as a tool in His continuing creation of the universe.

In our own times we have witnessed several brave religious pioneers who have marched into old areas of religious endeavor with a new, bold spirit and program. Brother Roger Schutz, the founder of the Taizé community in France, has answered one of the greatest spiritual needs of the post-war world. His quiet monastic community attracts architects, printers, theologians, lawyers, and countless professional people who, after submitting themselves to his program of prayer and reflection, return to the world to pursue their careers more fully committed to creating a more decent world of love and joy. His efforts to organize the worldwide "Council of Youth" in 1970 inspired thousands of young people to go to Taizé and then return to their own countries to work for religious renewal.

Mother Teresa of Calcutta, often called by people of many faiths "a living saint," has demonstrated to the world yet another way that God's creation can change through human (or in her case, superhuman) effort. Mother Teresa formed a new order of religious women that has lived among and served the poorest of the poor in India and many other nations, while at the same time demonstrating divine love. Public as well as private charitable organizations could follow her example and methods of providing human services and love to the outcasts of modern society. Malcolm Muggeridge said about Mother Teresa and her Missionaries of Charity:

When I think of them in Calcutta, as I often do, it is not the bare house in a dark slum that is conjured up in my mind, but a light shining and a joy abounding. I see them diligently and cheerfully constructing something beautiful for God out of the human misery and affliction that lies around them.[16]

Another pioneering woman in the struggle for Christian renewal in the world is Chiara Lubich. Her Focolare, or Fireside Movement, begun in Italy in 1943, has become a successful international means of providing spiritual community to people for whom the church as a system and institution is not enough. Living in lay communities structured as families, and imbued with the loving ethos of family life, architects, doctors, engineers, nurses, carpenters, secretaries, and others find a sense of spiritual belonging that run-of-the-mill society does not provide. Her innovative program, now adopted by thousands of people the world over, infuses vigorous inspiration into volunteers who seek to reanimate the world in the spirit of Christ.

Her New Family, New Humanity, and New Parishes movements are all creative changes in the traditional concept of
church organization.

It is through people like Brother Roger, Mother Teresa,
and Chiara Lubich that great blessings flow from the
church. More freedom should be given to people like these
three who take seriously the challenge to be humble co-creators with God. Their messages should be studied. The next
stage of human divine progress on the evolutionary scale
needs dedicated men and women, geniuses of the spirit,
blazing trails for the rest of us to follow. To encourage
progress of this kind, we have established the Templeton
Foundation Prizes for Progress in Religion. A list of award-
ees is included at the end of this book, in Appendix 6.

In recent centuries, hundreds of protestant denominations
have been born from new concepts and new revelations. Multitudes of cults and sects have arisen in other major religions,
also. But how many of these sponsor research for more new
ideas? The New Thought Movement, which includes The
Unity School of Christianity and The Church of Religious
Science, is a rare exception, one which strives for continuous
innovation.

Charles S. Braden writes,

New Thought as now taught is the creation of a perpetually ad-
vancing mind. It is not satisfied with any system originating in
other ages because systems do not grow while mind does. Indeed, change and growth are the silent mandates of Divinity.[17]

This is the kind of spirit the humble approach encourages, the
spirit that puts no limits on our quest for more understanding
of God. Elmer Gifford, a New Thought minister in Pasadena,
writes:

The term New Thought is used to convey the idea of an ever-
growing thought . . . man is an expanding idea in the mind of
God . . . . As mind advances, the old forms die, because they no
longer serve or satisfy men's needs . . . . New Thought can
never therefore be a finished product and if it remains truly
New Thought, it will never be completed enough to creedalize
it . . . . Thought can never be final and still remain thought.

The well-known scientist and Christian, Vannevar Bush, said, "A faith that is over-defined is the very faith most likely to prove inadequate to the great moments of life."[18] Certainly the great moments of life include those crises in which imaginative responses are needed. Marceline Bradford has said:

What is explicit here is the fact that millions of intellectuals the world over have become disenchanted with backward-looking religious institutions . . . . In order to recapture the great think-ing minds of the world, the clergy must turn their heads 180 degrees from past to future. With feet planted squarely in the present and eyes directed to the future, religious leaders can find factual bases in science for viable, solid, dynamic doctrines. For science and rationality are enemies not of religion—only of dogmatism.[19]

Even the best doctrines can affect some men like blinders on a horse. They can create a kind of tunnel vision. One unfortu-nate aspect of dogma is that it tends to belittle the infinite vari-ety and nature of God. Dogmas are, after all, written by men; and men's apprehension of God is always limited. A serious danger in this is that the great new advances and break-throughs in religion may come mainly from persons outside traditional church denominations. It would be unfortunate and a great waste of human energy if every advance or reform in the church had the nature of being a rival from the outside. Those well-meaning people within the church would naturally be more resistant, would tend to side with the conservatives, thus strengthening the forces behind the status quo and ulti-mately making change even harder to effect.

Most church concepts come directly or indirectly from an-cient scriptures. The problem with scriptures is that they were written in a world of men whose minds were limited by cos-mologies long since discredited. Today, we imagine the uni-verse to be billions of times larger and older and more complex than the one conceived by the ancients. The Bible, for ex-ample, is a most wonderful collection of God's revelations. But, we should seek to interpret it not as containing and constricting within its statements all that there is to be known about God, but rather as directing each generation of God's people to far,

far more of God than can ever be contained in the language and thought patterns of any age. To interpret the Bible in a narrow and restricted way, in accordance with the smallness of our mental grasp, is to make God smaller than our human minds. Two thousand years ago human minds were inevitably restricted to the range of their knowledge of the universe at that time, and they tended to confine their expression of divine revelation within that range. But, should we not be able to give a fuller and wider interpretation of divine revelation today, now that the range of our understanding of the universe God has created is so vastly enlarged by the discoveries God has allowed us to make? Why should we always try to express spiritual truths in obsolete words and ancient thought patterns? The fact that Jesus himself wrote nothing suggests that what he had to teach could not be frozen into words, even in his own age. Thus, he did not limit for future generations their range of human expression.

God has given us free will for new interpretations of eternal truths so that in our limited way we can be creative. But of course, free will gives us also the awesome power to build our own individual hells and heavens here on earth as well as for eternity. God knew free will would lead to problems, evils, and much suffering; but His divine plan seems to be that out of adversity we learn how to create; out of struggle we become more spiritual. Out of our desire to change the world for the better, we learn that the principle of creation is change and that through change will God's creations continue.

In conclusion, we might consider the description of reality given by Harold K. Schilling, a physicist:

I want to call attention to one more of the fundamental, constitutive characteristics of reality that provides, I feel, a criterion for the acceptability of values. Reality is historical and developmental, rather than merely inert and fixed; and it is creative and productive, rather than sterile and only conservative; and it is open, rather than closed, to new possibilities, and thrusts toward the future. Reality seems not to have come into being full-blown, but gradually, and over a long period of time.[20]

Schilling claims that matter and most likely all other forms of reality (forces? energies? spirits?) are fundamentally developmental. Reality is a continuing creative process in an unmistakable direction, "from the simple to the complex, from the small to the large, from the isolated individual entities to combinations and integrated systems, and to community." Links between people, between churches, and between nations need to be forged as a Mother Teresa or a Brother Roger or a Chiara Lubich would forge them. From the study of both physics and theology, the long-range cosmic trends seem obvious. In the words of Schilling:

In any case, it seems clear that if our values are to be in harmony with long-range trends—and divine intent—they must be such as not to hinder or inhibit development and change, and the emergence of the utterly novel, but to facilitate them, and thus to contribute to the building of a mankind and world characterized at its ground by development . . . .

Let us have no quarrel with any theologian. Let us happily admit that his concepts and doctrines may be right. But let us listen most carefully to any theologian who is humble enough to admit also that he may be wrong—or at least that his great insights do not close the door to great insights by others. Let us seek to learn from each other. Let us always keep trying to increase our humility.

VIII   Spiritual Progress

The idea of progress became a major goal in human thinking
only a few centuries ago in the West, and less than a century
ago in Asia and Africa. Men thought that the world (universe)
was static or moved in cycles, not in an upward curve called
progress. Therefore there was little if any urge towards scientific research. There was no concept of evolution for the earth,
for the universe, or for living things. Therefore men lacked the
urge to try by research to speed up or increase the course of
evolution. Until clear concepts of progress and evolution were
formulated in the eighteenth and nineteenth centuries, people
felt no need to use scientific research to understand the laws
of nature and society.

This static viewpoint still hinders most religions. Neither the
Koran, the Bible, nor the ancient scriptures of Asia say much
about progress, and they say even less about research. Even
now in the United States, the hotbed of research and progress,
church bodies do little to help in the evolution of religious
thought. They do almost nothing resembling the forward-look-
ing research being done in the great scientific laboratories of
corporations and universities. Within the different religious denominations of the church, the minor activities called research
are essentially archeological, concerned with the excavations of
ancient cities, the search for lost scriptures, or another modern
translation of an ancient book. In the United States alone over

250 billion dollars are spent yearly on scientific research, but almost nothing on spiritual research.

It is small wonder, then, that some people believe religion is gradually becoming obsolete. Mankind is racing forward in medicine and in genetics and in harnessing energy and understanding the nature of matter, but in spiritual growth the human race sometimes appears to be stuck in the stone or perhaps iron age. For thousands of years, numerous wise men and women have meditated and speculated about God. But so far there have been almost no experiments connected with spiritual realities in the same way that experiments are conducted by chemists and physiologists. Suppose chemistry were still dominated by alchemists searching ancient scriptures for lost secrets. What scientific breakthroughs would be produced by that thought pattern? Does this resemble in some ways the mental pattern of our churches and seminaries today?

We should listen to the warning implied in the anthropological studies of Anthony Wallace, who claims that in the last hundred thousand years of human history more than one hundred thousand different religions have flourished and disappeared. Wallace points out that this is surely evidence that in human nature there is always an abiding "sense of God." But, it is also a clear indication that concepts of God which are too small do become obsolete and vanish. History is replete with little gods who died. Herbert Menken compiled a list of over a hundred deities whose names now appear only in history books or as inscriptions on old monuments. Most likely they were much too finite to be gods at all. Belief in them died out because they were too statically or narrowly conceived; and they did not keep pace with the growth of men's minds, with the new knowledge, and the revelations the universal creative spirit continually gives to His children.

Often we read the words "religious revival" or "church renewal." Both are desirable; but they are not enough. What would it mean if people spoke in this way about astronomy, physics, chemistry, or medicine? Would "renewal of medical science" signify a lack of progress in the past needing new ideas? Would "revival of chemistry" imply a dying science in need of reviving or a need to restudy the ideas in the ancient books? To revive the old is not enough to keep theology in the vanguard of the knowledge explosion.

Religious thought should progress along with the social, political, economic, and scientific environment, or people will grow dissatisfied and abandon belief systems that appear to have little basis in reality. A *New York Times* study covering 1957 to 1970 chronicled Americans' replies to the question, "At the present time do you think religion as a whole is increasing its influence in American life or losing its influence?" The percentage who thought religion was indeed losing influence increased from fourteen percent in 1957 to seventy-five percent in 1970. Reasons given for this decline in religious influence included statements that religion was "outdated" or "not relevant in today's world." The report stated that these results revealed one of the most dramatic reversals in opinion in the history of polling.[21]

The Gallup organization has found even greater religious decline in Europe. Every nation in Europe has lower church attendance percentage-wise than in America. In some nations still considered heavily Christian, church attendance by adults now averages below ten percent. Such shrinkage appears in practically all the large, older Christian denominations. Also in the twentieth century, the percentage of practicing Buddhists and Hindus in the world population has decreased. As yet, the major denominations have rarely hired impartial scientists to study the causes of decline or the possible remedies.

Along with the decline in regular religious attendance at older chruches, hundreds of new religious groups have formed in recent times, and many of them have grown like wildfire in the last ten years. When the twentieth century began, for instance, there were no major denominations called "pentecostal" or "charismatic." Yet, beginning without any formal organization in dozens of nations, more than a hundred new denominations of these kinds have sprung up full of zeal and missionary outreach. In the United States alone, over fifty million people now describe themselves as "born again Christians." The first

World Congress of Charismatics drew over fifty thousand delegates to Kansas City in 1977 from all over the earth.

Television programs run by churches in the United States draw audiences ten times greater than similar programs of thirty years ago. At least a dozen church programs on television and radio enjoy audiences in the millions every week. More new church buildings are being erected than ever before. Circulation of church newspapers and magazines in the United States is breaking all-time records. In South Korea, the number of Christians doubled in the latest ten years and also in the ten years before that. Another doubling is expected in the next ten years.

Young people are responding with enthusiasm to new interdenominational and international youth clubs and organizations. Youth for Christ, Young Life, Inter-Varsity, and Campus Crusade have become very influential.

With this sudden upsurge in new religious activities, and with the need to keep religion relevant to the times, churches and foundations should now appropriate manpower and money for joint theological-scientific research. Governments should sponsor such impartial research just as they do science and cultural research. Because natural scientists understand the scientific method better than most theologians, they should undertake research projects independent of religious denominations and develop suitable methods of scientific inquiry into spiritual matters.

Recently scientific associations for this purpose are springing up, such as the Institute on Religion in an Age of Science and the Center for Advanced Study in Religion and Science, both at 1100 E. 55th St., Chicago, Illinois (60615-5199), and the American Scientific Affiliation at 55 Market St., Ipswich, Massachusetts (01938), and the Christian Medical and Dental Society, P.O. Box 830689, Richardson, Texas (75083-0689). Also, there is in Brussels the International Academy of Religious Sciences affiliated with the International Academy of the Philosophy of Sciences. Similar institutes concerned with science and theology have sprung up in recent

years in Heidelberg and Munich in Germany, and in Stras-
bourg and Metz in France, not to mention others.

Even more exciting is the vision of a new theology now being
born called the Theology of Science. Professor Ralph Wendell
Burhoe waxes poetic about this vision:

. . . it is still my bet that at several points in the next few years
and decades the traditional theological and religious communities will find the scientific revelations a gold mine, and that by
early in the third millennium A.D. a fantastic revitalization and
universalization of religion will sweep the world. The ecumeni-
cal power will come from a universalized and credible theology
and related religious practices, not from the politics of dying
institutions seeking strength in pooling their weaknesses . . . .

. . . I cannot imagine a more important bonanza for theologians and the future of religion than the information lode
revealed by the scientific community . . . . It provides us with a
clear connection between human values, including our highest
religious values, and the cosmic scheme of things.

. . . My prophecy, then, is that God talk, talk about the supreme determiner of human destiny, will in the next century
increasingly be fostered by the scientific community.[22]

Another striking phenomenon is the number of books on
theology being written by mathematicians, physicists, biologists,
and other natural scientists. Many are listed in the bibliography
of this book. Some scientists are reporting the results of their
research on the observable effects of divine activity in the
world. For example, Sir Alister Hardy, the marine biologist at
Manchester College, Oxford, is publishing a six-volume series
on the nature of religious experience. Journals, such as *Zygon*, a
quarterly published by Blackwells Publishers, Cambridge,
Massachusetts (02142), were founded to publish articles on
the relationship between science and religion. To paraphrase slightly William James, "Let empiricism once become
associated with religion . . . and I believe that a new era of
religion as well as of science will be ready to begin." Perhaps
it already has.

An age of "experimental theology" may be beginning. This term is used to indicate the study of unseen spiritual realities by concentrating on observable data resulting from spiritual events, changes, and differences in physical phenomena. Religious researchers should discuss ideas and propositions as openly as philosophers, and devise appropriate experiments to uncover new data about spiritual laws in the same way that scientists study the laws of nature. The great challenge at the present moment is to distinguish the real differences between the methodologies used to study the nature of physical phenomena and the methodologies to be used in inquiry into the nature of spiritual reality. The astronomer Carl Sagan wrote:

It was an astonishing insight by Albert Einstein, central to the theory of general relativity, that gravitation could be understood by setting the contracted Riemann-Christoffel tensor equal to zero. But this contention was accepted only because one could work out the detailed mathematical consequences of the equation, see where it made predictions different from those of Newtonian gravitation, and then turn to experiment to see which way Nature votes. In three remarkable experiments—the deflection of starlight when passing near the sun; the motion of the orbit of Mercury, the planet nearest to the sun; and the red shift of spectral lines in a strong stellar gravitational field—Nature voted for Einstein. But without these experimental tests, very few physicists would have accepted general relativity. There are many hypotheses in physics of almost comparable brilliance and elegance that have been rejected because they did not survive such a confrontation with experiment. In my view, the human condition would be greatly improved if such confrontations and willingness to reject hypotheses were a regular part of our social, political, economic, religious and cultural lives.[23]

It is generally acknowledged today in scientific circles that the so-called immutable laws of matter are merely descriptions of the way things usually happen. But every few years some scientist collects new data or proposes a new theory that requires some old law to be rewritten. We frequently hear of doctors proving that a certain treatment used in medicine a generation ago was worse than useless. In general, scientists are ready to test and accept new ideas that support the foundation of science and build on it, while repenting of inadequate theories and ideas that more recent evidence tends to discredit. A difference between scientists and theologians seems to be that whereas scientists and doctors are humbly admitting error, some theologians resent their pet ideas being challenged or criticized. In earlier, more barbaric days, religious authorities occasionally burned at the stake or crucified "upstarts" who claimed to have new revelations or alternative spiritual insights, even when they did not strike at the foundation of faith. Why should men of God be less willing to listen sympathetically to new ideas than scientists?

The possibility of a great new reformation depends upon scientists humble enough to admit that the unseen is vastly greater than the seen, and upon theologians humble enough to admit that some older concepts of God may need to grow. By using the humble approach, both can develop a vastly larger cosmology and a wider, deeper theology. Scientists, who are usually ready to have their theories rigorously tested, should undertake joint research projects with theologians in studying the relationship between science and religion. Their major contribution may be to show theologians the value of being open-minded, and the benefits that can be derived from a methodology that is willing to include new hypotheses rather than to excommunicate new hypothesizers. It would be a fabulous start if more religious leaders would encourage or merely allow their followers to read literature that challenges some older concepts. After all, a church whose doctrine cannot survive the fire of experimentation may lose its ability to set its members on fire with zeal and devotion. It may become as one of the dead churches which are now only a forgotten name in a dusty book.

No one can foresee exactly which research projects for spiritual progress should be undertaken or even the specific form that empirical inquiry may take in this realm. Nor can anyone foresee which experiments will prove fruitful. When research in electricity began in the eighteenth century, no one could have possibly predicted it would lead to telephones, X rays, and television. Research in the physical sciences shows that the great majority of projects result in no useful invention. Dr. Paul Ehrlich experimented with six-hundred-and-six chemicals before finding Salvarsan, the first cure for syphilis, called the "magic bullet." In natural science, research is largely a matter of trial and error. In empirical science it is only through multitudes of experiments that one new fact can be established.

No one can say in advance just what discoveries will be made by proper research in the science of the soul. Just as most medical experiments produce nothing immediately useful, very few empirical attempts to determine laws of the spirit will yield verifiable truth. Institutions willing to finance research can only accept project proposals and then select the few that seem to offer the best in terms of cost-benefit ratio. Where and how to undertake research will be learned from earlier research and from the free flow of information between researchers. This process of ongoing self-inquiry must become the normal method of spiritual studies since everything is so new and there are so few models to imitate. For example, experiments in natural science that produce the same results each time, like those in gravity, may not provide the paradigm for experiments appropriate to spirit. It is possible, however, that a statistical approach may yield limited results in a range of spiritual matters, just as sociological patterns are determined by studies of large groups of people or aggregates of similar events. For example, recent statistical experiments have convinced most people not that every smoker dies of lung cancer but rather that lung cancer is ten times more likely for cigarette smokers than other people.

At first, not all church dignitaries may welcome experimental questioning of this kind. They may not foresee the benefits from new discoveries in experimental theology. Many will find it easy to point out errors in the humble approach and this line of experimental inquiry, for there are sure to be multitudes of errors, as with all new projects. But if ninety-nine percent of spiritual experiments were failures and only one percent were to yield enduring benefits, this measure of success would be higher than that of the natural scientists.

Must it always be that resistance to progress is more often found in churches than among scientists or businessmen? Can breakthroughs in religion and great new advances come only from persons outside the hierarchies of church denominations? Why are more persons desiring progress attracted to careers in physiology or electronics rather than the ministry? Why are dictatorial "pronouncements" frequent in church councils but almost unknown in conventions of physicians or physicists? Is the basic cause a lack of humility? The smaller a man is the more likely he is to deny God altogether or to claim he knows all about God. Bigger men are usually more humble. God promises to reveal Himself to those who seek. If the seekers are physicists, then physics progresses more in its own realm than the church does in its realm. Can theology, once called the queen of the sciences, also use appropriate empirical methods of research? Can theology learn anything from the statistical methods of science? If that is possible, maybe it can become an adventure in the invention of research projects leading to the discovery and proof of facts about life in the spiritual universe. Likewise, researchers in spiritual matters should try out many different forms of inquiry and not be easily discouraged either. Researchers in relatively limited fields like biology have discovered only a little fraction of what may be known: hence, researchers about the Infinite could not expect to produce a systematic theology which is comprehensive and unchanging. With proper humility, they may hope to discover only a little more about man's relation to his Creator. Open-minded questioning and open-minded faith are quite similar. Both derive from humility. Both affirm that most is unknown and thereby keep the door open to further investigation and progress. Worshipers of every faith should seek and welcome opportunities to give proof and to determine the laws of the spiritual life which may convince the skeptics. Over six centuries ago, John Duns Scotus taught that the human intellect can know God through natural reason only up to a point, for knowledge about the Infinite cannot itself be infinite. Mere human intellect can discover a few spiritual truths without any miraculous illuminations, and even then it can take in relatively little of the living God and His ways.

As we have said, theological researchers must learn general theory as well as methodologies from scientists. Whatever their special field of interest, theologians should become to some extent theologians through science. Scientific revelations may be a gold mine for revitalizing religion in the twenty-first century.

The scientific concepts that appear inimical to religion at a primitive level of analysis may in the end contribute to the development of new universal symbols and languages that will keep human values and religious truths viable. Only by ongoing dialogue between theologians and scientists will the new patterns of culture emerge and preserve a sense of meaning to human life in the new age. Burhoe agrees with Anthony Wallace that religion

is the very center of man's most advanced evolutionary thrust to find order or organization, governing his overall attitudes and behaviors with respect not only to himself and his fellow men but also with regard to the ultimate realities of that cosmos in which he lives and moves and has his being.

In general, discussion about God in the next century will have to take place within the scientific community if it is to be widely relevant to the future society. It may be wise to begin those discussions now. Theologians following the humble way should now invite biologists, behaviorists, psychologists—scientists in all related fields—to offer courses in their seminaries and divinity schools.

However much we may agree or disagree with such a program and its predictions, it seems apparent that the scientific approach can rapidly produce a theology of such cosmic dimensions that it may resist the historic trend of obsolescence. Leaders of older religions should humbly recognize that God may be vastly greater than their earlier concepts of Him. Scientists should humbly recognize that they have gained very little insight into the nature of infinite God who creates universes both seen and unseen. God transcends and upholds all of nature, physical and spiritual. Nothing is separate from God. Human scholars participating in God's creative process can do so only by working in concert with the natural and spiritual laws of the universe.

The possibility that the earth may enter a new era of spirit beyond the noosphere is even more exciting than Burhoe's prediction regarding the theology of science. In fact, theology of science may be one step toward that era of spirit. As we have said earlier, the development of man on earth may not be the end of evolution, but only the beginning of it. It has always been difficult to imagine what would come next. However, the multitude of discoveries in this century related to things previously unseen, points toward the likelihood of even more amazing discoveries aiding human evolution in the future.

Sixty years ago, Professor Charles P. Steinmetz, director of General Electric Laboratories in Schenectady, said that when the great discoveries of the twentieth century go down in history, they will not be in natural science but in the realm of the spirit. Seventy years later we see little evidence of these discoveries. Could the reason be that we have poured more money and manpower into the natural sciences? Are we still blind to the possibility that the ever-new discoveries in natural sciences are actually data revealing the nature of the universal creative Spirit?

Let us try to devise some possible research projects in religion that might resemble current research in the physical world of science, medicine, economics, and politics. To serve as illustrations, here are a few various possibilities:

THE MINISTERIAL LONGEVITY PHENOMENON

Records kept for two hundred years by the Presbyterian Ministers' Fund, one of the oldest life insurance companies in the world, show that Christian ministers live longer than other men. Why? A research team of ministers, theologians, psychologists, and physicians might discover interesting information on this phenomenon. Has any scientist yet collected statistical data relating to whether medical doctors live longer than Christian Science practitioners?

HEALING AS MIRACLE

Several church denominations have collected thousands of well-documented cases of divine healing, but they have not yet been subjected to scientific studies by critically minded doctors, historians, and sociologists. Such studies may reveal how, why, when, and to whom divine healing most likely occurs.

THE RISE-UP-AND-WALK PROBLEM

Some doctors agree that the patients' rate of healing, after having the same operation, varies as greatly as three hundred percent among different people. In addition to studying the biological, anatomical, chemical, and psychological reasons for this, studies into the religious attitudes of patients might show a correlation between spiritual conviction and physical recovery.

THE JOY-TO-THE-WORLD QUESTION

St. Paul says that joy is one of the fruits of the spirit. Recent research has been conducted by psychologists on this question of why some people experience unexpected, intense rushes of joy while others do not. A theological consultant to these studies might be able to discover what *spiritual* factors contribute to the experience of joy. Are people who trust wholeheartedly in God generally more joyous than a control group of agnostics? Which groups of people describe themselves as happy most of the time? Which do not? Which groups radiate happiness? A scholar or researcher could raise interesting and pertinent issues on these questions that psychologists might otherwise overlook.

THE "BORN AGAIN" QUESTION

When people are "born again," or "filled with the spirit," they say they are no longer the same. Has anyone yet subjected such changes to rigorous testing? Scientists might be able to detect visible evidence about how the Holy Spirit alters or improves the lives of people who say they are filled with it. What characteristics "before and after" do these people exhibit? Maybe certain changes can be determined and described in science journals. What might be the outward aspects of the indwelling Holy Spirit?

PSYCHIATRIC HEALTH

Do persons who become charismatic Christians through the experience of Pentecost need psychiatric help less than they did before? Maybe scientists could collect statistics on the frequency of visits to psychiatrists before and after the charismatic experience. Studies of Catholics who see psychiatrists reveal that a great number of them no longer go to confession. Is there something about professing one's religion or confessing one's sins, i.e., discussing one's interior life openly with others, that eliminates the need for a psychiatrist's couch? Studies on this topic might be conducted.

GOD AND THE PSYCHOTIC

Psychiatric teams which include pastoral theologians might embark on new studies of patients in mental institutions to correlate types of insanity or psychoses with previous religious convictions. Are there significantly larger or smaller percentages of mental patients who were Christian ministers, atheists, Christian Science practitioners, doctors, nuns, or scientists?

THE PRODIGAL-SONS-AND-DAUGHTERS PROBLEM

It should not be left solely to sociologists, social workers, and parole officers to collect statistics on young criminal offenders in our society. Sunday schools, churches, mosques, and synagogues have been traditionally hailed as bulwarks of decent societies. Do youthful offenders come from families in which religious worship is strong, mediocre, or weak? Ministers and theologians should participate in these kinds of studies. It should be possible to collect statistics on young persons indicted for crimes to discover what proportion attended Sunday schools or were reared by parents who regularly attended church or mosque or synagogue.

FAITH AND HEALTH

Is it a fact that godly men have less heart trouble (or stomach trouble) than atheists? Has anyone bothered to make a careful investigation of this? Does any learned journal report such investigations?

These are but a few examples of the type of research projects in science and religion that institutes, academies, and seminaries might undertake. These examples are meant to be only thought-provoking. It is hoped they will suggest multitudes of other similar research projects.

Experiments with unquestionable scientific controls are convincing to educated leaders. Therefore people who love God should diligently work to devise thousands of appropriate, theologically and scientifically acceptable "experiments." Also, they should write articles for science journals and collect bibliographies of experimental research of this kind in the study of spiritual laws and observable effects of the unseen Spirit. Experiments by scientists who have earned the respect of other scientists can be especially influential. If the results of these inquiries prove to be convincing to college teachers, journalists, and government policy-makers, then the course of human history could be basically influenced towards bringing to all peoples "the fruit of the spirit."

To advocate experiments that would provide concrete evidence in religious issues is very different from advocating "free thought." The difference involves imagination and speculation versus critical testing and verification. Those who love God should participate in experiments related to spiritual matters and not leave them to researchers in psychic phenomena or extrasensory perception. The serious, conscientious researchers in those fields may be outnumbered by multitudes of the fuzzy-minded or opportunistic.

The better religious journals should publish articles discussing how far discoveries of astronomers and physicists may really be old revelations by God newly understood by man. Wherever truth is found, God is speaking. If the voice of science, however indirectly, is one of the voices of God, it should be listened to with reverence. If laws of the spirit are fragments of knowledge about God, so also in their way are the laws of nature. Churches and other religious groups should honor scientists who are committed to studying the divine as it manifests itself in the physical world. Seminaries should invite more scientists to lecture, and offer courses on topics related to theology through science, cosmology, and experimentation. We may be in for some surprising discoveries.

We may reaffirm what Henry Drummond discovered in the 1850s, namely that the supernatural is more natural than strange. He suggested that:

What is required to draw Science and Religion together again—for they began the centuries hand in hand—is the disclosure of the naturalness of the supernatural. Then, and not till then, will men see how true it is, that to be loyal to all of Nature, they must be loyal to the part defined as spiritual . . . . And even as the contribution of Science to Religion is the vindication of the naturalness of the Supernatural, so the gift of Religion to Science is the demonstration of the supernaturalness of the Natural. Thus, as the Supernatural become slowly Natural, will also the Natural become slowly Supernatural, until in the impersonal authority of Law men everywhere recognize the Authority of God.[24]

If Drummond was correct, we might find that members of both the scientific and religious communities have a common base from which to speak to each other.

Hence, the humble approach can be put into practice in a concrete way beginning now. The only requirement is that scholars in all fields recognize the mutual dependence and interrelationship of their disciplines. The next step is to bridge those mythical chasms that encourage experts in one field to talk only to other experts in that field. It may not be as difficult as some people have previously thought. Henry Margenau, professor of physics at Yale, said:

Science now acknowledges as real a host of entities that cannot be described completely in mechanistic or materialistic terms. For these reasons the demands which science makes upon religion when it examines religion's claims to truth have become distinctly more modest; the conflict between science and religion has become less sharp, and the strain of science upon religion has been greatly relieved. In fact, a situation seems to prevail in which the theologian can seriously listen to a scientist expounding his methodology with some expectation that the latter may ring a sympathetic chord. It is not altogether out of the question that the rules of scientific methodology are now sufficiently wide and flexible to embrace some forms of religion within the scientific domain.[25]

Mutual dialogue between studies of the natural and of the supernatural should begin in earnest, and we might put to rest once and for all those vague and damaging old claims that science has no soul and the soul has no basis in scientific fact.

IX The Benefits of Competition

The humble approach does not promise security; it is risky, strenuous, and dangerous. It will not be adopted by the weak of heart or will. Its advocates will be the strong-hearted and clear-minded, those who are unafraid of competition and struggle. Only those with a healthy faith in God and confidence in themselves will survive in this arena of ideas. The humble approach demands of its followers a dedication to freedom and a vigilant suspicion of security. As Benjamin Franklin put it in America's revolutionary days, "Anyone who gives up a little freedom for a little security does not deserve either."

The greatest development in human history has been the increasing possibility of each separate individual having the personal freedom to learn, to grow, and to design his or her own life. This is a relatively new phenomenon. Throughout recorded history most governments have been authoritarian, characterized by the rule of a few over the many. Such governments allowed little popular influence on decision-making in domestic laws or foreign policy, and little freedom of religion or speech. Most societies have not been "open" for the great majority of the populace.

Beginning around the time of the Renaissance, more emphasis was given to the expression of individual personality, to the notion that opportunity was a vital human concern, and to the idea that personal truth was more important that collective truth dictated from on high. Such notions led to the Reformation—a great breakthrough in religious freedom. The libertarian revolutions of the eighteenth century, coupled with the new scientific discoveries, produced the free world as we know it, a world where the individual is important, where a man's or woman's opinion should be listened to, where each person has a right to choose the basics of a decent life: career, marriage partner, school, religion, place of residence, and free speech.

Freedom fosters competition which yields progress. When the creativity and ingenuity and competition of individuals were set free, the result was progress and prosperity beyond anything ever imagined before. In the free world, amazing progress was made in the areas of education, religion, production, science, art, and literature. Inventions multiplied and culture was enriched. In God's ongoing creation, He seems to favor free, tolerant, open-minded individuals as His helpers. This individual freedom is enhanced by other freedoms—those of worship, free speech, free enterprise, and the right to own property. It is interesting that God does not force His good upon us. Instead He gives us the free will to claim or reject the blessings of life.

Only God knows whether the bounteous acceleration of progress and increase in knowledge over the past five hundred years will end. But if it does, most likely it will be halted by the greatest road-block to human progress and happiness: authoritarian government. Freedom requires government. As Jefferson put it, government is instituted to "secure" our liberties. But excessive government, the author of the Declaration of Independence warned, is the chief destroyer of freedom. Government maintains freedom to the extent that it gives each person freedom from domination by others, but unrestrained government intervention in the economic, cultural, or spiritual life of its citizens reduces that freedom. More research needs to be done on which government activities protect, and which restrict, freedom. Lincoln believed that government should do for its people those things that they could not do for themselves. But what are these activities?

Without going into detail, a democratic government by enacting clear and impartial laws usually protects freedom, whereas governments which respond to the whims of a dictator, an elite, or a bureaucracy lead to fear among the citizenry and domination of the people by the government instituted to serve them.

A most basic protection against bad government is the freedom to criticize and defy it. Those who can write or speak against those in government without fear of reprisal, are relatively free. But when the government is allowed to harass its citizens without due process of law, then freedom is a myth. Another safeguard of freedom is the right to ownership of property. People who have assets to own, use, and sell are much freer than those wholly dependent on assets owned by government or others.

No one knows for certain, but most people in the world are probably uneasy with the idea of individual freedom. In Erich Fromm's phrase, they seek an "escape from freedom." These people feel comfortable only under strong authority, as, for instance, when a dictator of some sort tells them exactly what to do and think. This is one reason why democracy has not proven to be a popular form of government in every nation. It is why many of America's Founders thought a democratic experiment would be short-lived. This history of democracies is that they don't last. Most people hunger for a demagogue to give them bread and circuses, fix their status, and give them a cog-role in the machinery of society. Such persons are only half alive. They seldom seek or achieve much progress, for progress comes only from the free, the pioneering innovators who take advantage of life's opportunities. It is difficult to imagine that God's creative purpose is served by those half-alive persons content to be only cogs in the regimented society.

Calvin held that people will instinctively worship, but if not taught to worship God they will create idols to worship. Today more people worship idols than God, and those idols are often the institutions and governments created by men themselves. A good example was found in the communist world. Eight hundred million Chinese communists venerated Mao Tse - tung as if he were a god. For many years more copies of this "little red book" were sold than either the Bible or the Koran. In general, communist dedication and devotion to government resembles religion. Demanding total surrender and total loyalty, communism, like a political religion, formulates elaborate and inflexible doctrines to control every aspect of life. It resembles slavery. It may claim to be revolutionary but in actual fact it is reactionary, a regress to the kind of serfdom typical of the Dark Ages when government and church dictated the details of each individual's daily life.

No one knows why God allows his people to fall into captivity, as has happened so often in history. Perhaps He allowed communism to establish a place in people's minds and hearts because of the vacuum created by religions that failed to develop and progress with the "knowledge explosion." Maybe the communist threat served the purpose of challenging the increasing religious and social rigidities that occur when there are no great challenges to face. Adversity does seem to serve a creative purpose. An omnipotent God might permit dictators to stamp out religion for hundreds of millions of people—for an ultimately useful purpose: to awaken and challenge the faithful. It may be a "call to arms" for His children to take a stand against tyranny and fight for freedom so that progress may continue. Tyranny of any kind hinders the great creative process of evolution.

Communism, too, is becoming obsolete, because after the first revolutions, its doctrines became rigid and forbade change. In the rise of any new ideology, rigid doctrine leads to the triumph of rigidity itself. Then it causes decay and obsolescence. Decay became evident in many communist countries because their governments prevented freedom in so many areas of personal life—economic, religious, intellectual, and artistic. Being so earth-centered, communism became as narrow-minded as the cosmologies of the ancients who pictured the universe extending for only a few thousand miles and enclosed within the star-studded spheres of heaven.

Communism was founded on hatred between the classes and envy of those to whom God gives more talents. So it was unlikely to uplift human nature. Communism substituted the collective structure of government for the individual as the giver of welfare to the poor. But the greater blessing from giving goes not to the receiver but to the loving giver. Jesus said, "It is more blessed to give than to receive." (Acts 20:35 KJV.) The giver is the one who is uplifted and spiritualized by the gift. When giving is channeled through government by force, it has an opposite effect on the producer whose produce is forcibly given. Forced charity does not uplift either the giver or the receiver. It dehumanizes and hinders the spiritual growth of all the people. Forced charity may be better than no charity, but it is a pitiful substitute. Never did Jesus advocate government welfare! Never did the disciples or prophets of other religions either, except those who were earthly dictators themselves.

Based on man's vices rather than virtues, communism did not uplift human nature. All in all, communism seems to be one of those worldviews out of step with our modern notion of progress, a notoin that includes personal freedom and a humble acceptance of scientific data suggesting a spiritual realm beyond material reality.

People who like "all-pervasive" government seem to be attracted to careers in government offices, whether of churches or nations. Perhaps this explains why so many in church headquarters advocate expansion of government regulations. Moreover, advocates of policies leading to more regimentation seem to form multitudes of lobbying organizations; whereas those who like more freedom for each individual are less likely to get organized. It is strange how rarely church leaders speak out today against dictatorial governments or atheistic regimes. Why are they so silent on great moral issues, such as personal freedom?

Rather than advocate more individual freedom, some religious leaders, particularly the bureaucratic-minded ones, support a union of all Christian denominations. At first, union sounds desirable. Such advocates say it is a scandal that Christ's churches are divided among themselves. One slogan often heard is that "united we stand but divided we fall." But should reconciliation in Christ mean the institution of a centralized authority?

Some leaders once hoped that the World Council of Churches would lead to a world union of churches. Many writers, especially some Hindus, advocate a union of all religions. We should make every effort to tolerate each other, listen to each other, and eventually love each other. Many benefits flow from sharing literature and information with each other and from facilities to help us lovingly listen to each other.

But centralized authority from a bureaucracy of united world churches is the prelude to stagnation. Religion will grow not in union but through freedom and competition. Originality and discovery derive from variety, not uniformity. If each denomination of every faith sent missionaries to every other denomination lovingly to explain and share their vision with others, then we would all learn more. If our concepts are divine truths, then they will not suffer in competition with other concepts. What a happy competition it will be if each of us is lovingly trying to give to the other his holy treasures. How selfish it would be for us not to witness to others and send missionaries abroad! But likewise, other religions or denominations want lovingly to witness to us and we should receive their unselfish missionaries with gratitude and brotherly joy.

Among scientists there is friendly rivalry between those holding different theories and those belonging to different schools of thought; and if churches do not foster something similar, they will appear old-fashioned and then irrelevant. God's long creative process has always led to greater diversity. Let us rejoice in diversity and avoid establishing any monolithic church. By spirited and loving competition, the truth will be purified and strengthened.

Progress comes from competition and this is what churches need most. Competition between religions may be an advantage. By free competition the wheat is gradually separated from the chaff. The true religions should welcome competition because then they are put to the test, and if they are true, they will survive. Only an inferior religion needs to prevent competition, lest its inferiority should be exposed. The long history of the evolution of plant and animal varieties seems to show that competition is God's chosen method of developing his creation. Why should it be different in the realm of spirit and religion?

Tolerance is a divine virtue but can become a vehicle for apathy. Millions of people are thoroughly tolerant toward diverse religions; but rarely do such people go down in history as creators or benefactors or leaders of any religion. The use of tolerance is mainly to keep us humble so that we may listen with an open heart and an inquiring mind. We want our neighbor to share and try to convey to us the brilliant light which transformed his life—the fire in his soul—not his least-common denominator. More than tolerance, we need competition. When men on fire for a great gospel compete lovingly to give their finest treasures to each other, then everyone benefits. If we enthusiastically share the inspiring highlights of each faith and church, then we are all richer. On our daily upward struggle toward communion and union with our Creator, we need all the inspirations and revelations which every child of God can give us.

In summarizing the views of Teilhard, Wildiers said there are two ways of unifying mankind, by coercion and by sympathy and affection. The truly creative force is a combination of sympathy and affection. Even at lower levels than the human, these forces operate to unify and create the universe. They are

the constructive forces in the cosmos as a whole. The atoms were impelled towards one another by an intrinsic affinity; and so the molecules came into being. The cells coalesced; and thence the great diversity of organisms appeared.[26]

Another area in life where the law of competition is being undermined is in the lack of teaching of religion in our schools. In recent years American schools have come to teach mainly secular rather than religious knowledge. No wonder misbehavior is increasing. Where are the schools and colleges which still consider it their first duty to teach honesty, ethics, character, morals, self-control, philosophy of life, and religion? Most churches seem content to devote only one hour a week in Sunday school to teach our children principles governing spiritual growth. Many Sunday schools rely only on amateur teachers and require no homework, no grades, and no tests. How woefully ignorant our children would be if we relied on such methods to teach history or science! It is no wonder that we are said to be rearing generations of moral and spiritual weaklings.

Until the last century, most colleges and universities were founded and supported by religious groups. Often the main purpose was to prepare priests and ministers for their lifework. Most schools for children were founded and taught by religious people. Now, most of these schools and colleges, in the name of impartiality, have abandoned responsibility for teaching religion in favor of secular subjects. The desire for impartiality does not mean we should eliminate religion from the curriculum.

In the name of equality or social justice, governments increasingly dominate educational policies and squeeze out the church-supported schools. In the United States, the courts rule today that the Constitution forbids religion in public schools, even though the authors of the Constitution had no such intent. The men who framed the Constitution of the United States would be dumbfounded by the Supreme Court decision that forbade worship in public schools. The founders of our nation intended to ensure free and fair competition between religions, not to stamp out religion. Their efforts to separate church and state were not efforts to abolish religious education in the classroom. Total omission of religion from schools prevents free competition. The net effect is to imply that only secular, not spiritual knowledge is respectable. New generations are being educated to be intellectual and cultural adults but spiritual and ethical infants.

Various methods have been devised to allow each child, or its parent, to choose from the broad spectrum of religious studies—from traditional world religions to atheism. There are ways to keep religion in our schools without favoring one denomination over the others. For example, each school could provide a room which could be used for thirty minutes a day by any religious or nonreligious group to present their beliefs or worship. Students or their parents could then choose which room they will study in.

Competition is God's law of the universe because through it the useless are gradually weeded out and change for the better occurs. God's law of evolution means progress. If we do not have free competition among ideas, even religious ideas, old and established theories will never be improved. They will never have to defend themselves. They may survive long after their time of usefulness. We should not abolish the discussion of religion in the classroom anymore than we should ban the discussion of political or biological theories.

We should not even ban worship. Rather, let us have different forms of worship in schools and allow our children to inspect and experience them all. Those which are most meaningfull will be well-attended. Those that students find worthless will eventually be discontinued. If we did this, there would be healthy and lively rivalry among various denominations and intense discussion about the Spirit. Some students might even discover their future careers as ministers or theologians or religious scholars because of exposure to religion at the lower levels of education. This is the humble approach: an open mind, a willingness to admit there might be alternative truths just as valuable as ours, and the fortitude to compete with others—to search, to discover and create the societies and religions of the future.

Another technique for providing free public education and religious instruction for those who desire it, would be for the government to compute its cost of educating each child for a year and then give to parents a voucher with which they could pay tuition costs at any school of their choice. The selected school would then redeem the voucher for cash from the government. Not only would such a system provide both free education *and* religious education; it would also introduce competition between schools which would lead to better teaching methods. Recently, certain churches in many nations have founded and financed church schools for the children of their members. At present, parents who want their children to learn character and religion in schools must pay twice; first in cash to the church school and then in compulsory taxes to the government for schools to which they will not send their children.

The wisdom of God is beyond the measure of men's minds. To most of us, the "survival of the fittest" seems cruel. God allows big fish to eat little fish, and he also allows millions of insects to die every hour. For many people, the idea that "survival of the fittest" builds stronger species does not justify the suffering. However, maybe God in his infinite wisdom and mercy so regulates the course of His creation that the suffering involved in this harsh law is the best way to build and enrich souls.

We should remind ourselves once more that to embark upon the humble approach is to ask for competition in the battle of ideas. To assure that the competition is fair, no church, no school, no government should seek to impose its system upon others who do not share their beliefs. Church-goers, students, and citizens need absolute freedom so that unfettered they might examine the rich variety of ideas in God's universe. We need to be constantly vigilant against any person, group, institution, or political party that would tyrannize our lives. To assure free competition, Thomas Jefferson proposed "eternal enmity over every form of tyranny over the mind of man."

X Earth as a School

What is the purpose of life on earth? Many philosophers have said that God created the earth as a school for souls. Some say human souls are created out of the thoughts of God. God may have invented and created each body and mind as the temporary shell to house a soul during its childhood.

Rev. Earnest O. Martin said in 1962:

. . . spiritual development, the formation of a heavenly character. This is the purpose of life on this planet, which Swedenborg spoke of as a seminary for heaven. It is here that we grow into angelhood and begin to develop the potentialities that God sees within us. Heaven is essentially a quality of life, an inward condition or state in which men live in harmony with the will of God. And yet it is also a place—a real, tangible, substantial existence. Swedenborg's answer is that life is one. The natural world and the spiritual world are not two distinct, separate existences that have no relationship. We are spirits and from the day of our conception we are citizens of the spiritual world. Love, understanding, loyalty, friendship, patience, mercy,— these are spiritual realities the Lord seeks to instill in our lives here and now. The natural world is the theater in which our spirits operate and develop and grow. It is here that our loves, our attitudes and our desires are molded and find expression. That is why God has placed us here.

An omnipotent God could have created us all as angels instantly. But, He did not choose that method. A lifetime on earth may seem a slow way to create a soul. And eighteen billion years may seem a slow way to create the school building; but, let us remember that God also created time. Nels F. S. Ferré (1908–1972) used to remind us that God is not bound by time, for He created not only our universe but also both space and time. Time may have been created as the means and medium for learning, for growth, and for building the ability to give love. William Adams Brown (1865–1906) taught that the purpose of God in creating the world is to develop the Kingdom of God, or more specifically:

. . . the production beings like the good God, and their union with Himself in the fellowship of life. . . . it is because the world as we know it today ministers to such a spiritual end that we believe it had its origin in the will of the holy and loving Father whom Christ reveals.[27]

The great question is this: How much progress can our soul make before our body becomes uninhabitable? To progress is to increase our love of God, our understanding of God, and our love for His children. Our body has a physical reality, but it is only a temporary shell. Death destroys only that which is fit for destruction. The butterfly developing in the chrysalis in due time splits and abandons the dead chrysalis and flies away on wings of amazing beauty undreamed of by the chrysalis or the caterpillar.

Various major religions have described earth as a school. The *Bhagavad-Gita* ("Song of God") teaches:

Whatever a man remembers at the last, when he is leaving the body, will be realized by him in the hereafter, because that will be what his mind has most constantly dwelt on during this life. Therefore you must remember me at all times and do your duty. If your mind and heart are set upon me constantly, you will come to me. Never doubt this . . . I am the Atman that dwells in the heart of every mortal creature; I am the beginning, the lifespan, and the end of all . . . one atom of myself sustains the universe.

Buddhism, teaching that the life of the spirit transcends the life of a man on earth, emphasizes that this life of the spirit is the only true reality and that on earth people should strive to grow spiritually through exercising free will, reason, love, and meditation.

Rev. Charles Neal, in 1976, expressed it in this way:

"Who then are we?" We are God's perfect children in the making. Each of us is evolving to perfection. But at this stage we do not yet have it made. Let us recognise that the Universe is a Cosmos. It is not slaphappy, disorganized, haphazard or accidental. It is an orderly system and its nature is to evolve. That is, to bring into the open through endless and infinitely slow degrees that which is already involved in it, namely Perfect Creation. God-mind moved on itself; It got an idea, to express Itself. It went to work, and out of the formless it created form. How did it all start? The "big bang" theory is as good as any. Possibly eons ago all space was filled with whirling gasses, minute particles of matter. Then drawn together by some force (love?) until they became compressed generating heat, then exploded into suns and stars. Eventually cooling through millions of years and forming various layers: the central core, the barysphere; the crust or lithosphere; then the waters forming the hydrosphere; then the atmosphere and finally the stratosphere. . . .

The nature of the Universe is to evolve. This is true of every part including rocks, minerals, plants, animals and man. This is also true of the very essence of the Universe, which is consciousness. Consciousness has been present in every stage. It is God-Mind seeking to express Itself as perfection. Consciousness evolves along with every other element. A rock is more limited in consciousness than the mineral it contains. A vegetable with its ability to grow is a still more highly evolved expression of consciousness. As the Persian poet wrote: "God sleeps in the mineral, dreams in the vegetable, stirs in the animal, and awakens in man."[28]

Earlier in the century, Teilhard, speaking from long years as a scientist, priest, and poet, said, "It is a law of the universe that in all things there is prior existence. Before every form there is a prior, but lesser evolved form. Each one of us is evolving towards the God-head." This evolving toward God may be our purpose on earth.

If earth is a school, who are the teachers? One teacher is called adversity. Why did God put souls into a world of tribulations? Why did He not just make souls perfect in the first place? Is not God vastly more far-sighted and infinitely wiser than we are? Maybe from God's perspective the sorrows and tribulations of this earth are the best way to educate souls.

Growth can come through trial and self-discipline. There is a wealth of evidence indicating that too much prosperity without work weakens character and causes us to become self-centered rather than God-centered. Spiritual growth and happiness do not come from getting but from learning to give. The great souls are the most rapidly growing souls. Trees and human bodies are limited in growth both in space and time, but is there any evidence that the individual soul is limited in its growth?

How could a soul understand divine joy or be thankful for heaven if it had not previously experienced earth? How could a soul comprehend the joy of surrender to God's will, if it had never witnessed the hell men make on earth by trying to rely on self-will or to rely on another frail human or on a soulless man-made government?

Maybe the earth was designed as a place of hardship because it is the best way to build a soul—the best way to teach spiritual joy versus the bodily ills. Why was it said that into every life some rain must fall? It is apparent that sometimes a great soul does not develop until that person has gone through some great tragedy. Let us humbly admit that God knows best how to build a soul. If the soul were born perfect, how would it understand or appreciate the absence of pain and sorrow? As a good father does not do his son's homework for him, so our Heavenly Father does what helps us to grow, not what we ask for. We should be thankful that God does not always give us the stupid things we ask for. St. Paul wrote:

More than this: let us even exult in our present sufferings, because we know that suffering trains us to endure, and endurance brings proof that we have stood the test, and this proof is the ground of hope. Such a hope is no mockery, because God's love has flooded our inmost heart through the Holy Spirit he has given us. (Romans 5:3-5 NEB.)

As a furnace purifies gold, so may life purify souls. When a man is born into the world, he is like a piece of charcoal. It is soft and amorphous so when rays of the sun fall upon it, it reflects nothing. Then in the crucible it is subjected to such intense pressure and heat that it is born again as a diamond. Next, it is cut with many facets by the master craftsman. Now when the sun's rays fall on it, it reflects the colors of the rainbow, creating a symphony of beauty and radiance. So it is with a man between the time he is born into the material world where he is cut and chipped and then born again into heaven when he begins to reflect the divine light of God. Maybe this was God's purpose for creating the crucible called earth.

How does a soul grow? Spiritual growth takes place through human reason and divine revelation, through communing with nature and God, and by diligent use of the talents He gives us as well as from learning how to make studied and wise choices. Man's capacity for choosing good over evil must be developed. In exercising our free will and choosing good, we demonstrate that we are created in the image of God. Each time we consciously prefer evil over good, we have failed another test in becoming more God-like. By being a positive creator in nature, rather than a destroyer, we may become more and more like the infinite Creator.

Just as politics is essentially control over others, so is religion essentially control over, or overcoming of, self. Those who have learned to put aside completely their selfish egos become as a little child and enter into the kingdom of heaven. Our job is to be pure like a clean windowpane so that the truth and light of God can shine through—a radio receiver through which God's music enters the world. When we know this, we become constantly grateful to God and good things are attracted to us as to a magnet. St. Paul wrote, "For we are God's handiwork, created in Christ Jesus to devote ourselves to the good deeds, for which God has designed us." (Ephesians 8:10, NEB.)

There are many similarities between the Exodus story of forty years' wandering by the chosen people in the Wilderness of Sin before reaching the promised land "flowing with milk and honey," and the forty years wandering by each person between birth and the time when some learn the spiritual laws and reach heaven on earth. Often it takes forty years for a self-centered little animal to become a God-centered little angel. What are the marks of spiritual maturity? Do we put our trust in God or in man? Do we desire more to give than to get? These are marks of a mature soul.

Spiritual growth can be achieved in part by knowledge—by overcoming our ignorance and self-centeredness until we are in tune with the divine. Since God possesses all knowledge, any additional knowledge or understanding we can acquire makes us more like Him. As followers of the humble approach, we must always be alert to new discoveries and new insights into both spiritual and natural phenomena. We should never imagine we know enough. That would be like saying we know God in his totality, which would border on the blasphemous. Ralph Waldo Trine put it this way:

Great spiritual truths—truths of the real life—are the same in all ages, and will come to any man and any woman who will make the conditions whereby they can come. God speaks wherever He finds a humble listening ear, whether it be Jew or Gentile, Hindu or Parsee, American or East Indian, Christian or Bushman. It is the realm of the inner life that we should wisely give more attention to. The springs of life are all from within. We must make the right mental condition, and we must couple with it faith and expectancy. We should also give sufficient time in the quiet, that we may clearly hear and rightly interpret.[29]

Of course all of us should work for self-improvement by prayer, worship, study, and meditation. But one of the laws of the spirit seems to be that self-improvement comes mainly from trying to help others—especially from trying to help others to enjoy spiritual growth. Growth comes by humbly seeking to be a more useful tool in God's hands. Giving to others material things helps the growth of the giver, but often injures the receiver. It is better to help the receiver to find ways to grow spiritually himself. It is more far-sighted to give advice and instruction, like a wise father to the son whom he loves. If, following Jesus, we teach "Seek ye first the Kingdom of God," then the other material things will follow. Helping the poor to grow spiritually and to become givers themselves is the real road to permanent riches, including material riches. Over and over it has been proved that there is magic in tithing.

Charles Grandison Finney, the great revivalist of the nineteenth century, taught that insofar as people can be brought to obey God, economic and social troubles will fade away. After a person is born again, he or she is twice as useful in the world, helping to bring salvation to others. Finney's great belief was that for every opportunity missed, a soul is lost. Hence the lives of Christians ought to be lived at fever-pitch like those of a rescue party during a disaster. We should not sing, "The great church victorious will be the church at rest."[30] This is like saying graduate from the university so you can stop learning. Spiritual growth does not result from rest. If the earth is a great classroom for souls, troubles and strife are the examinations. God counts not whether we live or die, but how we meet the tests. We receive blessings in proportion as we use the talents and blessings already given us. But the work most often neglected is the work of spiritual education. If God created the earth as a school for souls, then most of us are very lazy pupils. Most of us may fail the test. Life on earth is brief. We are in this school only a few years. Why should we waste even one day? Each night when we lie down to sleep can we say we have learned to love God more or helped our neighbor to love God more?

One of the major lessons to learn while on earth is that building our heaven is up to us. Emanuel Swedenborg wrote that we will not be in heaven until heaven is in us. Here on earth we can begin to receive the life and spirit of heaven within us. Swedenborg also believed that heaven is a kingdom of "uses," where everyone is challenged to his or her full potential, where contributing to the welfare of others brings happiness. Heaven is glimpsed on the day we realize that each of us has a unique and valuable gift to benefit the world.

In the gospels, of course, there are two differing views of the kingdom of God. The apostles themselves did not seem to ever get it clear. One is the external rule of God predicted as a purely future event. The other is the internal rule of God already present here and now. Jesus spoke more often about this internal condition whereby a person is born again and his or her whole existence is permeated by the living God. It is this concept of the kingdom that makes more sense for those adhering to the humble approach.

More and more theologians agree that heaven and hell should be spiritually interpreted, "as conditions in consciousness and not as geographic locations" or "areas of awareness in which the soul or psyche or spirit continues its development nearer to the reality of God than (is) possible on this four-dimensional physical plane." This was Charles Fillmore's view. "The 'many mansions' are no longer etheric apartment dwellings, but, according to modern theology, states of mind or of being."

The concept that heaven only comes after death or is located elsewhere fails to encourage us to build heaven on earth. However, if we view the earth as God's garden for nurturing souls, then we should follow Jesus' example of witnessing to that kingdom of God on earth now. In each miracle of healing, Jesus did not say, "I will heal you after you die or when I come again." He said, "Rise up now and walk." In and through Jesus' example, God helps people to grow spiritually. Such teachings are important for showing us more of the nature of God. They instruct us in ways to become more God-like, and assure us we can reach heaven during this life as well as after death.

The idea that heaven is a locality situated on the other side of death and that we must be either wholly in heaven or wholly in hell, has stunted progress in religion and caused it to march out of step with the rapid progress of business and science. Should we not see all around us that heaven is like sanity or wisdom? Few are totally without it. It usually comes gradually, not all at once. Some travel back and forth between heaven and hell. Those who are most deeply in heaven radiate happiness, but their circumstances vary greatly in health, wealth, and intellect. Heaven may be in the union of our spirit with God's spirit, in a constant striving and studying to become like Him. When Jesus said that His kingdom was not of this world, He may have meant that it is within our mind and soul, not in outward material surroundings. As St. Paul told the Corinthians, "Well, now is the favorable time; this is the day of salvation." (II Corinthians 6:20 JB.) St. Luke 17:20–21, KJV, reads: "And when he was demanded of the Pharisees, when the kingdom of God should come, he answered them and said, The kingdom of God cometh not with observation. Neither shall they say, Lo here! or, lo there! for, behold, the kingdom of God is within you." Luke 13:18–21 tells us that Christ said: " 'What is the kingdom of God like?' He continued, 'what shall I compare it with? It is like a mustard seed which a man took and sowed in his garden; and it grew to be a tree and the birds come to roost among its branches.' Again he said, 'The kingdom of God, what shall I compare it with? It is like yeast which a woman took and mixed with half a hundred-weight of flour till it was all leavened.' " How should we understand what Christ said in Matthew 16:28 (KJV)? "Verily, I say unto you. There be some standing here, which shall not taste death, till they see the Son of Man coming in His kingdom."

If a man's soul grows until it reaches heaven on earth, then for him individually there has been a "second coming of Christ"—for him individually "the end of the world" has come and gone. He or she is "in the world but not of it." Much of the eschatological prophecy of the Bible can be interpreted in this spiritual way to apply inwardly to the individual human soul rather than to earthly locations or to the end of humankind as a whole. We who advocate the humble approach should be open to the possibility of reaching heaven on earth. Standing for spiritual progress, as it does, the humble approach is always "in" this world but not "of" it. It is always moving into the next world or the next phase of human development.

Who are the happiest people you have ever met? Let us write down the names of ten persons who continually bubble over with happiness, and we will probably find that most are men and women who radiate love for everyone. They are happy deep inside themselves because they are growing spiritually and fulfilling God's laws. Jesus said: "Thou shalt love the Lord thy God with all thy heart, and with all thy soul and with all thy mind. This is the first and great commandment. And the second is like unto it, Thou shalt love thy neighbor as thyself."

Each human can only study and strive toward such divine love. Even the saints need to work daily to maintain continuous overflowing love for friends and foe alike. But the more we love, the easier it becomes to love even more. Love given multiplies. Love hoarded disappears.

This is a law of love. If we radiate love, we will receive back joy, prosperity, happiness, peace, and long life. But if we give love only to gain one of these rewards, then we have not understood love. Love in expectation of any reward is not love. When we learn to radiate love, we are fulfilling God's purpose— bringing God into expression on earth. When we learn to radiate love, we are gaining that mind which was in Christ, and are becoming fruitful sons of God, opening the door to heaven on earth.

Thus, through personal adversity, service to others, study of the natural and supernatural cosmos, prayer, and meditation, we can advance in the spiritual journey. We will find, however, that the journey does not lead to some atmospheric cloud-land of harps and haloes. It leads within. The search for God, as Carl Jung has pointed out, will take us through our own unconscious into the kingdom of heaven that is at hand, the state of perfection that lies in embryo within each of us.

From the variety of nations on earth, would we find it more pleasant to live in a nation of atheists or a nation of theists? What differences could we expect to find? From any community, select the dozen most beloved men or women. Will they be mostly God's men? Because they radiate love, love returns to them twofold. The reality of God is shown by the gifts He leaves with those who open their hearts to Him, gifts such as joy, mercy, and love. This distinguishes spiritual communion from hallucination and superstition which rarely leave such gifts. Let us think over the people we know to see which have come closest to heaven on earth. Are these mostly God-centered persons? One way to prove the existence of God is to see what results for those who put their faith in Him.

Samuel Taylor Coleridge (1772–1834) said that the truth of

religion is found in the way it answers the deepest needs of a man's spirit. In America it has been estimated that one out of every twenty adults suffers from alcoholism, and more in some other countries. But among the fifty million Americans who now claim to have been born again, alcoholism is very rare. God cures them when they turn to Him. This is a blessing for their families also. When we "get religion" even our dog or cat can see a change in us.

XI Creative Thinking

Frequently people say, "I'm going out of my mind" or "my mind is playing tricks on me" or "I'm losing my mind." The mind is certainly an elusive component of every human being. It is the link between our bodies and our souls. With it we can think about our nature, other people, our bodies, and our souls. We can even think about thinking; that is, we can use our minds to think about our minds. Most likely, this is what distinguishes our mental capacity from that of animals. Animals are at a less evolved level than man precisely because they do not seem to be able to step back mentally and think about themselves. Animals have instincts, some level of primitive thought, and even emotion. They can build environments and communicate with each other, even with us. But as far as we can tell, animals cannot philosophize about anything.

As human beings we have been given by God a highly developed, extremely sophisticated mind. According to His plan, evolution has produced us for a purpose: possibly to be co-creators with Him in His on-going creation of the cosmos. He endowed us with the mental capacity to know, learn, evaluate, solve problems, and improve our selves and our societies. In this way we are created somewhat in His own image and likeness. With minds attuned and receptive to His revelations, both spiritual and natural, we can be a help in future progress.

God has given each person a mind capable of creative activity in the maturation of the universe, as well as the maturation of the soul. By following the humble approach, we will keep our minds as open and receptive as possible because we never know what contingencies await us. We should also keep our minds strongly linked to our souls and our souls linked to God. In this way, the creative process in which we are engaged will flow from the mind of God through our souls to our minds where creative thinking will produce creative results in the physical world.

We will see the divine effecting changes in the external culture we humans create within our homes, families, schools, churches, businesses, and governments. We will also be aware of spiritual evolution in our own personalities. On the other hand, if we let slip our control over our thoughts, we might produce disastrous effects, much like persons who think or feel their way psychosomatically into illness. Our minds are powerful. They can bring on physical sickness and determine the rate of recovery. If they can unconsciously produce such results within the cover of our skin, imagine what they could achieve with conscious effort in our exterior world!

Many people, however, are as self-centered as little children. Gradually they learn to love more selflessly and thus reach spiritual maturity. Such persons begin their entry into heaven while on earth. They can make the most of their years on earth, to develop their spiritual talents, as well as to serve the needs of others. For example, each person can learn in many ways to radiate love. One of the simplest ways is by thanksgiving. A heart full of gratitude is ready for love. If, before arising, we say a prayer of thanksgiving each morning—naming five of our blessings—we are ready that day to radiate love. God surrounds us each day with so many thousands of blessings and miracles in His universe, that it is easy to start each day with deep thanksgiving.

Since we have the gift of self-consciousness, we should examine our lives from time to time to see how we are using our mind, imagination, and will to build God's kingdom through love. One way is by creative control of our thoughts. Each word or action begins as a thought. We can say loving words and do loving deeds only if our minds and hearts are full of love. So it is good to stand watch over each thought which sprouts in the mind. If it is a loving thought, let us cultivate it. If it is not, let us crowd it out by filling our mind with loving thoughts. Try this experiment: Think of any person you envy or resent. Then free the mind of that poison by seeing that person as a child of God, and pray for that person's happiness as an experiment in self-discipline. The God of Creation dwells in various ways in every human being, and it is easy to find good qualities in any person if we lovingly try. As Jesus said, we should pray for those who irritate us. Our mind is very much like a garden, fertile with good soil, water, sunlight, and drainage. We, as the gardeners of our minds, can cultivate whatever thoughts we choose; we must nourish good thoughts, weed out the bad ones, and ensure that evil thinking does not overshadow and blockout the radiance of the good. By years of careful thought control, your mind can become a garden of indescribable beauty. "As a man thinketh so is he."

For the sake of clarity and simplicity, let us think of each person as composed of four basic components: *God, soul, mind,* and *body.* Philosophers, cosmologists, theologians, and poets have used a great variety of words, synonyms, and analogies to describe these distinctive realities. But the multitude of terms tends to obscure exact meanings. Of course this quartet is an oversimplification. Reality is vastly more complex and infinite.

However, for the sake of simplicity, we will consider *God* to be the infinite creator of the cosmos. If He is truly infinite, then nothing exists apart from Him, and all other realities are created reflections of Him. *Soul* signifies that divine infusion which is unique to each human being. Most major religions teach that each soul is immortal and can be educated. *Mind* is defined as the strategic link between soul and body. Mind is complex and miraculous but temporary. The human *body* is defined as a temporary physical dwelling for the mind and soul. In the light of scientific analysis, all bodies, human and subhuman, are only evanescent arrangements of forces and wave patterns.

Obviously the human body, especially the brain, has great effect on the mind. When we are tired or sick, for instance, we often think less clearly. But basically the determining influence flows in the other direction, from the unseen to the seen, from God to soul, to mind, to body. These four are realities; but probably the lesser three derive from the first.

When we look at a person, we see a body and recognize a mind. When God looks at a person he may see first and foremost an immortal soul. We should never forget that the body is only a temporary configuration of wave patterns. First appearances seldom reveal the underlying truths. The earth appears to be flat but actually is round. It appears to be still but is actually moving in various directions at incredible speeds. The egg looks dead and inert until the day the bird breaks out. As Emerson said, "Things we see are out-picturing more basic realities that we do not see." The ultimate causes, forces, and purposes of things are invisible. We usually see only a few outward manifestations, not the causes and forces themselves, but only their effects.

Our body is like our house. So is our mind. Let us take care of it while we live in it, using our mind to develop our soul. God gave it to us not in fee but as a temporary trust. Why worry if our mind is feeble, if it is good enough to connect our soul with the world while we develop our soul? The life lived in the open is only a surface reflection of the life lived in closed sessions with our God. The laws which govern the realm of soul can be learned, but not always in the way we learn the laws of arithmetic or physics or human logic.

Ramakrishna once said:

It is the mind that makes one wise or ignorant, bound or emancipated. One is holy because of his mind, one is wicked because of his mind, one is a sinner because of his mind and it is the mind that makes one virtuous. So he whose mind is always fixed on God requires no other practices, devotion or spiritual exercises.

Unfortunately, most of us do not possess the Hindu claim that allows total concentration on God. We need to work at controlling our minds and channeling our thoughts. Norman Vincent Peale's approach—"positive thinking"—is a good one to follow.

Ella Wheeler Wilcox (1855–1919) wrote in her poem, "Attainment":

Use all your hidden forces. Do not miss the purpose
of this life, and do not wait for circumstances to mold
or change your fate!

And, in "You Never Can Tell", she wrote:

"You never can tell what your thoughts will do in bringing
  you hate or love,
For thoughts are things, and their airey wings
Are swifter than carrier doves
They follow the law of the Universe,
Each thing must create its kind,
And they speed o'er the track to bring you back
Whatever went out of your mind."

Marcus Bach wrote:

. . . to dwell upon goodness is to become the recipient of all
that is good. To be joyful is to attract more joyfulness.
To meditate on love is to grow lovely in the inner self.
To think health is to be healthy. To put life into God's
love is to put God's love into your life.

There is an intimate relationship between what we think and
what we are. The Dhammapada scriptures, attributed to the
Buddha, put it this way:

All that we are is the result of what we have thought: it is
founded on our thoughts, it is made up of our thoughts. If a
man speaks or acts with an evil thought, pain follows him, as
the wheel follows the foot of the ox that draws the carriage.
. . . If a man speaks or acts with a pure thought, happiness
follows him, like a shadow that never leaves him. . . . Hatred
does not cease by hatred at any time; hatred ceases by love
—this is an old rule. . . . As rain breaks through an ill-
thatched house, passion will break through an unreflecting
mind. . . . If one man conquer in battle a thousand times a
thousand men, and another conquer himself, he is the greatest
of conquerors. One's own self conquered is better than all
other people conquered; not even a god could change into de-
feat the victory of a man who vanquished himself.

As co-creators with God, we must control our thoughts, directing them toward creativity, toward evolution, toward progress. If our thoughts are not in this direction, we may be impediments in God's plan.

Why do not all people learn to benefit from mind-control which is probably the most useful tool we will ever possess. Some people allow their minds to run as loose and as uncontrolled as a spilled cup of mercury. Some are not even aware that they can control their minds. Some are just too lazy. In fact, the mind is only an instrument. Each person can guide thoughts and even emotions into any paths he wishes. We can each learn to be the pilot of our mind. A ship without a pilot leads to disaster. So does a mind and body without a pilot. Even ordinary people wield extraordinary power in the words that leave their lips. Jesus was addressing each of us when he said:

For the words that the mouth utters come from the overflow of the heart. A good man produces good from the store of good within himself; and an evil man from evil within, produces evil. I tell you this: there is not a thoughtless word that comes from men's lips but they will have to account for it on the day of judgement. For out of your own mouth you will be acquitted; out of your own mouth you will be condemned. (Matthew 12:35–37, NEB.)

God has given each person a mind capable of creative activity in the ongoing expansion of the cosmos, which includes the expansion of his own soul. The creative process goes from thoughts to deeds. Our words are our thoughts crystallized. The objects we build and the deeds we do emanate from our thoughts and our words.

Obviously, to build a house we begin with thoughts, then words, then deeds. Every good object produced by man is created by this process. Nations are formed in this way, and so are sciences and all the organizations and institutions of human society. Even more awesome is the fact that thoughts build not only outwardly but inwardly. By thoughts we create not only our possessions but also our personalities and our souls. By long practice of thought control we can each make of ourselves the kind of person we want to be. Our thoughts and attitudes can even help to bring about physical sickness, but they can also influence the rate of healing in our bodies as well as our minds.

It cannot be repeated often enough that we can learn to radiate love; but first we must practice using loving words and loving thoughts. If we keep our minds filled with good thoughts of love, giving, and thanksgiving, they will spill over into our words and deeds. If we are not very careful to weed out all evil thoughts such as envy or hate or selfishness they, too, will overflow into our words and deeds. To produce beautiful music requires long practice and so does the production of a beautiful mind. With practice both come more easily. We must beware of entertaining whatever pops into our minds. Rather, let us think thoughts we may be proud of, controlling and using them as divinely provided tools. If we really do not want to worry, then we must fill our minds with thoughts of thanksgiving, not fear. This takes self-control and practice; but the more we practice, the easier it is.

Charles Neal said in 1978:

One would think therefore that intelligent men and women, particularly prayerful ones, would be meticulous in their choice of words, speaking only those which are helpful and are constructive. And above all things, let us avoid negative and destructive words. But as we know very well, this is not what usually happens, for people still do not really believe that they may have dominion over their own lives, and fail to exercise that dominion through the right use of their minds and through the right use of selected words. While their minds have creative power, people do not seem to realize this, despite the fact that it is clearly implied again and again in our scriptures. . . . Our casual words reflect what we truly believe deep down within ourselves. They express what our thoughts have been over long periods of time and they come through spontaneously, unimpeded by deliberate and considered thought. Our so-called thoughtless words are reckoned to be true expressions of what actually is in our minds. Thus we begin to see that what some people call slips of the tongue or Freudian slips are not slips at all. Detail of sickness, disharmony and woe, that every counsellor and New Thought minister hears over and over again, are the products of minds that have made wrong choices in the use of ideas. Surely the cure for sickness, disharmony and woe of this kind lies in a complete change in the disciplined management of our thought process, including the very words that we utter, day by day.[31]

Our thoughts and words have power. We should discipline and manage them wisely. They can be what separate us from the rest of creation. They can also be what connect us to God.

XII Love and Happiness
The True Test

Happiness sought eludes. Happiness given returns. The pursuit of happiness is never successful because happiness is always a by-product. Ralph Waldo Trine (1866–1936) has said:

A corollary of the great principle already enunciated might be formulated thus: there is no such thing as finding true happiness by searching for it directly. It must come, if it come at all, indirectly, or by the service, the love and the happiness we give to others. So there is no such thing as finding true greatness by searching for it directly. It always, without a single exception, has come indirectly in this same way, and it is not at all probable that this great eternal law is going to be changed to suit any particular case or cases. Then recognize it, put your life into harmony with it, and reap the rewards of its observance, or fail to recognize it and pay the penalty accordingly; for the law itself will remain unchanged. Life is not, we may say, for mere passing pleasure, but for the highest unfoldment that one can attain to, the noblest character that one can grow and for the greatest service that one can render to all mankind. In this, however, we will find the highest pleasure, for in this the only real pleasure lies.[32]

If we endorse the humble approach, we should radiate love and happiness as faithfully as the sun radiates light and warmth. As sunlight is a creative source, so can our love be a creative source of new life and ideas. God is the source of love. Love cannot flow in unless it also flows out. The Spirit of God is like a stream of water and His disciples are like many beautiful fountains fed by this river of waters. Each one of us is such a fountain, and it is our task to keep the channel open so that God's Spirit can flow through us and others can see His glory. Without God, we are not likely to bring forth any good. If we think too much of the visible world or trust in our own ability, we become like a clogged fountain. We will never learn to radiate love as long as we love ourselves, for if we are characterized by self-concern, we will radiate self-concern.

Jesus then said to His disciples "If anyone wishes to be a follower of mine, he must leave self behind; he must take up his cross and come with me. Whoever cares for his own safety is lost; but if a man will let himself be lost for my sake, he will find his true self. (St. Matthew 16:24–5, NEB.)

God loves us all equally and unceasingly. It is His nature to do so. We should seek always to let God's love shine forth like the light inside an electric bulb illuminating all our habitation. Emil Brunner wrote:

Every human relationship which does not express love is abnormal. In Jesus Christ we are told that this love is the whole meaning of our life, and is also its foundation. Here the Creator reveals Himself as the One who has created us in love, by love, for love. He reveals to us our true nature, and He gives it back to us.[33]

Rufus Matthew Jones taught that those who worship God are empowered by the Spirit, and that religion is not a burden but rather a matter of being lifted up to new heights of joy and philanthropic achievement. The divine spirit moves into your life and makes it over from within so that all things are seen in a new light, and love for all becomes the spontaneous expression of a Spirit-filled soul. If we think too much of the visible world, or trust too much in our own ability, or love ourselves inordinately, we may never learn to radiate love. Instead we will radiate self-concern, egotism, arrogance. Ironically, egotism will not allow us to find our true selves. We miss the true self of grace and spirit, and thus sever our links with God. The self we admire so greatly is a diminished version of the total self. Often in the name of humanism, egotism obscures the divine life in our souls. It substitutes the worship of men for the worship of God. In its extreme form, godless humanism becomes idolatry, denying the existence of the infinite creator and exalting the limited intellect of the human race so recently evolved on one small planet, earth.

The ultimate conceit is that no mind is greater than the human mind. How could humans be so egotistical as to believe that no other being in the universe can be purposeful and consciously creative? Or that creatures on other planets may not have learned more about God than we? Surely man's multitudinous creations are tiny and few compared to what the mind which created man has created in other galaxies and in unseen realms and dimensions undreamed of.

Humanists claim they promote science and progress. However, we must not immediately assume that they espouse the humble approach. In his recent work on humanism, *Architects of Moral Anarchy,* John Howard has discussed the dangers to society when traditional moral precepts, such as hard work and respect for the property of others, are discarded for an ethic of selfishness. He decries the moral vacuum created in the last forty years in which the structure of a hard-working society is ridiculed. The work ethic and other virtues of free competition may not survive in a moral vacuum. To prosper, society needs citizens deeply convinced of the value of earning one's way from "rags to respectability." Howard writes:

Nowadays, anyone who ventures into the territory of moral judgments, declaring certain conduct right or wrong, is simply ignored, or in the case of an influential spokesman, will have scorn heaped upon him by the public commentators. When a Solzhenitsyn describes with chilling accuracy the moral rot that has debilitated America, he must be chastised, belittled, and declared incompetent to judge such things.[34]

Before remedial action can be undertaken, the forces causing this situation should be clearly understood. Possibly the leading force in the so-called "new morality" lies in the twentieth-century humanist movement as expressed in The Humanist Manifesto, proclaimed by thirty-four respected Americans in 1933. (The present humanist movement, Howard warns, is not to be mistaken for its historical forerunner which dates back to Renaissance humanism. Erasmus, Melanchthon, Luther, and More were poles apart from modern humanists. The crucial difference, of course, is faith in God.) In Howard's analysis, the 1933 document

asserted that science and economic change had rendered old beliefs invalid and that conscientious people needed to recognize that new circumstances required a formulation of a value system appropriate to the new knowledge. The cornerstone of the new doctrine was to be an acknowledgement that there is no Supreme Being, with the obligations, hopes and taboos of traditional religion set aside. Man was declared the measure of all things, so there needed to be a hierarchy of commitments to give human beings the best chance for satisfaction and joy in life. In short, a new humanistic religion was declared.

Howard maintains that these modern humanists convinced themselves that a socialized, cooperative order should be established for the sake of redistributing more equitably the means of life. He said:

The moral fervor of this movement and its categorical rejection of traditional beliefs made humanism the seedbed of, and the rallying point for, many of the new thinkers in education, psychology and the other social sciences, as well as the activist leadership in most of the civil rights and liberation movements.

For four decades, ideologies stripped of God provided a sufficient formal agenda for the energetic and increasingly influential humanist movement. Then in 1973, *Humanist Manifesto II* was published in *The Humanist,* the official journal of the American Humanist Association. Atheism was reaffirmed, and traditional religion identified more specifically as an obstacle to the complete flowering of the human psyche. Ethics were proclaimed to be "autonomous and situational," and traditional codes of morality were decried. A guaranteed annual income was requested. There were endorsements for an international redistribution of wealth, pacifism, and a world government.

In 1969, Dr. Carl Rogers, officially designated "Humanist of the Year," gave the Sonoma College commencement address entitled "The Person of Tomorrow." In it, Rogers said the person of tomorrow has no use for religion and marriage as institution. He is "passionate" and "spontaneous" but for no concrete, humanitarian goal. He trusts only his own experience, obeys only the laws he chooses, and exhibits "a profound distrust of all authority." It appears that this person is the epitome of self-centeredness. By definition, humanism is man-centered and therefore egotistic.

Such a person could hardly be a follower of the humble approach that encourages us to use our open-mindedness and our inventiveness to advance God's evolving creation. This "person of tomorrow," admired by Rogers and others, is too egotistical, too dogmatic, too selfish to achieve his stated goal (which, again, is far removed from that of the humble approach) "bringing to all the people of the world the maximum joy and satisfaction as defined by the scientific knowledge of the moment." But there is more than the knowledge of the moment. Unfortunately, a self-directed, situational ethic admits of no eternal truth, no infinite divine creator, no enduring obligations.

The degree to which this philosophy has been accepted and woven into the institutions, the life-styles, and the expectations of society will determine the size of the task confronting anyone seeking to re-establish the habit of viewing the world in moral terms and thus preserving the system of competition, freedom, and advancement through healthy rivalry. Whereas many humanists have very genuinely believed that the new morality to which they are committed constitutes a higher set of virtues than that which has prevailed in the past, the editor of *The Humanist* resorted to sensationalism by placing an advertisement for that journal in *Hustler* magazine. Bragging that, "For more than three decades *The Humanist* has chased almost every sacred cow in the country," the ad claimed that *The Humanist* is "one of the most irreverent, most quoted and most influential magazines of social commentary in America."

How much respect can one have for a movement that demeans the long line of sincere humanists stretching back to the great names of the Renaissance by using a pornographic magazine as an invitation to others to "learn more about us and our jousts with sacred truths"?

The humble approach rejects all self-centered philosophies, especially those brands of humanism teaching that man is the end purpose of evolution. In humility let us admit that God's awesome creative process is likely to continue even if humans should disappear from the face of the earth. Humanism is egotistical because it encourages men to think that mankind is, itself, the ultimate concern. Rather than worshiping the universal Spirit creating the universe, humanists worship the creatures. Communists go even further to worship governments which are created by creatures. Today we wonder why the children of Israel in the wilderness worshiped a golden calf which they themselves had created. Is it any wiser to worship a government created by men?

Many people in this generation—including some intellectuals and academics attracted to humanism—have turned away from God because of the traditionalism of so many churches. They think of the church authorities as old-fashioned or even small-minded. But is not that very turning away equally small-minded?

The humble search for God continues in new ways, and seeks to recreate and revitalize the churches. Have we ever met a person who was helped to grow spiritually by turning away from all religion? Do those who turn away become better persons, more useful, or more high-minded? Would it not be more beneficial for humankind, if persons to whom God has given keen intellects use their talents for increasing our understanding and love of God?

Rejecting religion is negative and self-centered. Even the disillusioned could become positive, generous forces for reform if they did not retreat into their private worlds of pessimism and despair, like Jean Paul Richter, Friedrich Nietzsche, and Albert Camus, the trio who laid the foundation for the current idea that "God is dead." If we accept such a ridiculous notion that God died, we must then accept the notion that there is no more revelation, no new guidelines; that man is now on his own as he moves into the next era of change. Those adhering to the humble approach cannot accept a position that replaces the primary active source of all knowledge, God, with weak human intellect. The humble approach includes a vision of progress as limitless, unending—a vision that calls us to do our best work, to serve others, to love, to aspire beyond the merely human.

By their fruits you will know them: the egotists, the humanists, the immoralists, those who despair. And by the fruits of faith, hope, and love, you will know the humble children of God. Let us explore the ways that these traditional virtues should be reflected in those who follow the humble approach.

Faith

University students often give up their faith mistakenly thinking "faith" is similar to a kind of blindness. Some become self-styled atheists believing in only the visible or tangible. Later, after much more experience in life, most of them turn back to religion. Life and experience teach that things visible to humans are ultimately unexplainable and unknowable without an understanding of underlying spiritual realities. The know-it-all attitude is often found among the young. As people grow older, however, most acquire a deeper faith in God and more religious fervor. Egotistical arrogance often comes from a lack of experience, and humility comes with maturity. The gullibility and credulousness of some young persons is not at all the same as the broadmindedness and humility of the humble approach.

Faith does not imply a closed, but an open mind. Faith means "having respect for" or "standing in awe of." Quite the opposite of blindness, faith appreciates the vast spiritual realities that materialists overlook by getting trapped in the purely physical. Where the Bible says "fear of God," possibly a better rendering in many passages would be "respect for God" or "admiration for the great power of God." Gabriel Marcel claimed that "an open and expectant state of mind . . . either implies faith or is faith."

How does scripture define faith? In the New Testament we learn that "Faith gives substance to our hopes and makes us certain of realities we do not see." (Hebrews 11:1, NEB.) The faith which Jesus requested of us differed in some ways from what was later called faith by the church. Jesus taught personal loyalty or faithfulness. The church later sometimes seemed to mean something more like credulity. Mere credulity ought to be replaced by a conception of faith as an attitude of humility and loyalty which results in expectancy and bestows a quiet confidence.

Hope

Basically, hope is trust. In what or whom do we trust? The humble approach bases its hope on the eternal, not the transient. Understand the transient and utilize it, but do not place eternal hope in it. Can there ever be any question more basic than this: "Do we put our trust in God or in some temporary thing?" God will never fail us. A father may fail us. A son may fail us. Possessions may fail us. But God's love never fails. God's love is always unconditional and unlimited even for the worst sinner. If we put our trust (faith) in God, we are opening the door for abundant blessings.

The man who trusts in himself, thereby cuts himself off from God and from other men. Trust in self (egotism) leads to worry, tension, and even to insanity. No man can solve all the problems which sometimes come into a human life. Subconsciously he knows he is not equal to the task and may crack under the strain. But to trust in God does not mean that we should not work hard and use to the utmost each of our talents. Jesus made this duty plain in the parable of the talents. A man who is not productive becomes unhappy. To trust in God (faith) means that we should remember that each of our talents is a gift from God. If we are strong in mind or in muscle, that, too, is a gift from God. If we have honors or titles or possessions, they, too, are God's gifts. It is a form of idolatry (and egotism) to put our trust in any of these. Our knowledge may be famous among men; but compared to God's knowledge, it is infinitesimal. Therefore, the truly wise and humble man prays, "Not my will but thine be done." Trust in others weakens; trust in self isolates; but trust in God magnifies our life and our soul.

Charity

What is charity according to the humble approach? It is not the same as welfare. Giving welfare to the unemployed, although good for the soul of the giver, is not good for the soul of the receiver. The greater charity is to help a person join the ranks of the givers rather than be forever trapped in the position of receiver. When the desire to give replaces the desire to get, a person exchanges an attitude of childishness for one of maturity. Just as when one nation gives food to another nation, it does no spiritual good for the recipient nation. In fact, human nature is such that the receivers often become envious, demanding, and resentful. This is bad for the souls of people.

The greater charity is to help the people of a poor nation to learn the traits which lead to prosperity, such as trustworthiness, hard work, free competition, thrift, and the Golden Rule. Any poor nation can be converted to amazing prosperity rather quickly and lastingly if its people and government adopt wholeheartedly these five virtues. Not only is this the way to material prosperity but also to spiritual growth. The godly gift to the poor is not money but the gospel.

The wealth of a nation comes not from natural resources but from what is in the minds and hearts of its people. People pursuing the humble approach express their love in charity in order to both alleviate suffering and elevate the recipients of their love. Feeding the hungry, caring for the sick, clothing the naked are necessary in the short range; but in the long range, the real charity is to help the poor learn the spiritual traits which lead to prosperity, dignity, and happiness. Besides bestowing technology and know-how on people in poor nations, we should also support the self-sacrificing missionaries who live among the less fortunate as loving human helpers in their development. Technology and know-how are spiritless. People, however, can radiate love and joy as they teach the spiritual realities which are creating the material universe.

XIII   Laws of the Spirit

Everyone now understands the term, "the laws of nature," the multitude of principles discovered by scientists, primarily in the last four centuries, explaining or describing the physical universe. Our faith in their immutability has been shaken in recent years as we see them being revised, reformulated, and replaced because of new data. Nevertheless, we now know that such laws exist and from time to time we experience their effects in our daily lives.

Not everyone, however, yet understands the phrase, "the laws of the spirit." There is a difference between laws of the spirit and religious laws such as those formulated by Moses, Hammurabi, Mohammed, and other ancient law-givers. Early civilizations adopted complex and rather rigid systems of religious laws. Some of these religious teachings may have reflected underlying laws of the spirit, but most were not themselves laws of the spirit. Some tended to legitimize social and religious customs of ancient cultures. By "laws of the spirit" we mean universal principles of the unseen world that can be determined and tested by extensive examination of human behavior and other data. Partly for lack of clearly defined methodology and a body of research material, this field appears about as disorganized and controversial as the natural sciences were in the milleniums before Galileo. In the days of Moses or Mohammed, there was very little knowledge of the principles of physics, chemistry, or biology, and little appreciation among the average people of the progress and rich rewards that could be achieved through successful research in these fields.

Just as people of earlier times were ignorant regarding the physical sciences, we today are ignorant about the principles of spiritual progress. In addition to having as yet little understanding or agreement as to what spiritual laws are, we do not seem to recognize that God's purpose is not some permanent status quo, but change, process, and progress based upon the laws of the spirit. The spiritual dimensions of the cosmos are dynamic, changing, ceaselessly interacting. Surely the time has come for us to concentrate our resources on the kind of investigations which will enable us to understand the patterns and laws governing spiritual growth and development. Ceaseless seeking may be a part of the growth of souls as well as minds. It may be possible through research that some agreement may be reached on laws of the spirit. This field of research may yet become as bounteously fruitful as the natural sciences were in the last four centuries.

Church leaders often are not yet aware of the need to sponsor wide-ranging open-minded research into the laws of the spirit. History records innumerable crises in which some traditional leaders of established churches did not welcome innovative suggestions for religious renewal and advancement. Many accused these "researchers" of being heretics, self-seekers, or insane. Indeed, some were, just as today some studies in psychic research are conducted by fuzzy-minded or poorly trained psuedo-scientists. If a new renaissance is to begin now, research into laws of the spirit should be undertaken largely by persons rigorously trained in many fields of science: physical, mental, and spiritual. Already, some leaders in these exact sciences are beginning to speak and write about their religious concepts. We hope they will help to clarify or verify some of the laws of the spirit, as in recent years some medical investigations into ancient herbal remedies have proved many of them to be useful.

We should first become comfortable with the concept of spiritual laws and not fear that research into them violates some static condition of God's universe. We may indeed discover that some laws are eternal verities, never to be altered. But with an open-minded humble approach, let us begin. The Book of Psalms opens with these words:

"Blessed is the man . . . (whose) delight is in the law of the Lord; and in his law doth he meditate day and night and he shall be like a tree planted by streams of water, that bringeth forth its fruit in its season, and whose leaf doth not wither; and in whatsoever he doeth he shall prosper." (Psalms 1:1–3, KJV.)

Of course, it is not very apparent at this moment in history exactly how each law of the spirit could be discovered, tested, and utilized. Nor can we predict what laws will be discovered by generations of scientists sifting data for evidence of the effects of the unseen. It would have been impossible five hundred years ago to predict anything of the laws of thermodynamics or nuclear physics, let alone the devising of experiments to test and establish them as laws. The difficulties of bringing to light, of describing and testing the laws of the spirit, are not any less than those which faced chemists two hundred years ago.

Studying and teaching the laws of the spirit should benefit humanity in even greater measure than did, for example, the laws of chemistry. Matthew Arnold thought that the decreasing influence of the Bible in the nineteenth century could be reversed if the ideals and hopes and laws expressed in the poetic and allegorical language of the scriptures could be explored experimentally. He hoped that dogmatic theology could be succeeded by empirical theology or experimental theology. If people could understand religious principles in their own everyday language rather than in ancient metaphors, they might take them more seriously. Likewise, if we could see the effects of spiritual laws operating in our modern world rather than read about their effects in ancient Israel, we might try to live our lives in closer harmony with them.

Possibly there are some laws of the spirit on which all major religions agree? For example, ancient Hindus taught that hate is never overcome by hate, but hate can be overcome by love. Does any religion now disagree? Such agreements might be useful in providing a starting point for further inquiry, experimental testing, and formulation.

It is not necessary, however, or even desirable that everyone should agree on a fixed code or list of laws. Natural scientists are not in absolute agreement about laws of nature. Some diversity of judgment and opinion can be beneficial because that induces progress. Maybe the amazing acceleration of discoveries and progress in the twentieth century is due in part to increasing diversity and rivalry in the domains of the sciences. If a world government were to codify a list of laws, perhaps progress and advance in research would slow down dramatically.

More benefits may result in the domains of the spirit if each individual were to draw up his own personal list of the laws governing spiritual matters. Of course, this would be easier if he first studied the books and articles of scientists engaged in investigating possible laws of the spirit. However, only when we understand and claim as our own some actual laws of the spirit do we begin to build our own heaven. What could be more uplifting than for each human to write in his mind and heart, as well as on paper, the various laws by which he ought to live? He may measure his spiritual growth if every year he revises and rewrites his own personal list of laws. How beneficial it would be if every school each day devoted a few minutes to help each pupil study the laws of the spirit as they are brought to light and formulated by great scientists, so that each person could improve his own written list. The supreme moments in the life of each of us occur whenever we grasp a new inspiring truth and appropriate it so that it revitalizes our personality and becomes a part of our life.

When any field of research is begun, no one can possibly predict what may be discovered eventually. Astronomers before Copernicus could not have predicted or even imagined galaxies, or supernovas, or pulsars. Even so, no one can yet say what laws of the spirit will be formulated and proven eventually. But to give some idea of laws of the spirit which might be researched, here are a few examples taken mostly from the wisdom of the centuries expressed by the world's major religions.

1. Religions generally agree that, "As a person thinks, so is that person." If this is a true law and it can be taught con-

vincingly, especially to young people, it might be the basis for new generations much more disciplined in the control and management of their minds and lives than current generations.

2. Happiness comes from spiritual wealth, not material wealth. Happiness is always a by-product, never a product. Happiness comes from giving, not getting. If we pursue happiness for ourselves, it will always elude us. If we try hard to bring happiness to others, we cannot stop it from coming to us also. The more we try to give it away, the more it comes back to us multiplied. If we try to grasp happiness, it always escapes us; if we try to hand it out to others, it sticks to our hands like glue.

3. The more love we give away, the more we have left. The laws of love differ from the laws of arithmetic. Love hoarded dwindles, but love given grows. If we give all our love, we will have more left than he who saves some. Giving love, not receiving, is important; but when we give with no thought of receiving, we automatically and inescapably receive abundantly. Heaven is a by-product of love. When we say "I love you," we mean that "a little of God's love flows from us to you." But, thereby, we do not love less, but more. For in flowing the quantity is magnified. God's love is infinite, and is directed equally to each person, but it seems to gain intensity when directed to sinners. This is the wonder and mystery of it, that when we love God we get an enormous increase in the quantity flowing through us to others.

4. It is better to give than to receive. Giving is a sign of psychological and spiritual maturity. There are few diseases so childish and so deadly as the "gimmies," a disease that separates us from friends and from God, and shrinks the soul. The secret of success is giving, not getting. To get joy we must give it and to keep joy we must scatter it. The greatest charity is to help a person change from being a receiver to being a giver.

5. Loneliness is the punishment for those who want to get, not give. Helping others is the cure for loneliness. If we feel lonely, we are probably self-centered. If we feel unloved, we are probably unloving. If we love only ourselves, we may be the only persons to love us. Whatever we give out, we get back.

6. Thanksgiving opens the door to spiritual growth. If there is any day in our life which is not thanksgiving day, then we are not fully alive. Counting our blessings attracts blessings. Counting our blessings each morning starts a day full of blessings. Thanksgiving brings God's bounty. From gratitude comes riches—from complaints poverty. Thankfulness opens the door to happiness. Thanksgiving causes giving. Thanksgiving puts our mind in tune with the Infinite. Continual gratitude dissolves our worries.

7. To be forgiven, we must first forgive. Forgiving brings forgiveness. Failure to forgive creates a hell for the unforgiver, not the unforgiven.

8. When Jesus was asked what is the greatest law, He said:

Thou shalt love the Lord thy God with all thy heart, and with all thy soul, and with all thy mind. This is the first and great commandment. And the second is like unto it, Thou shalt love thy neighbour as thyself. On these two commandments hang all the law and the prophets. (Matthew 22:37–40 KJV.)

This can be researched as a basic law of the spirit. A person who applies this law finds his life revolutionized. Opening our heart to God allows His love to flow through us like a mighty river. If we love God totally as He loves us, we will love each of His children without exception as Jesus Himself described. (Luke 6:27–36, KJV.) The happiest people on earth are those who love God totally.

9. Surrender to God brings freedom. It is in dying to our selfish selves (self-denial) that we are born to eternal life.

10. St. Matthew reports that Jesus enunciated another law of the spirit: "But seek ye first His kingdom and His righteousness, and all these things shall be yours as well." (St. Matthew 6:33, RSV.) Do we not have daily evidence of this in the confirmation of the proverb, "Honesty is the best policy"? Honesty leads to prosperity. Those who are trustworthy are the ones entrusted with great blessings. Even in the lives of nations this law of the spirit is manifest, for any nation whose people are steadfast in religion, fair play, hard work, thrift, and trustworthiness will automatically grow in prosperity also.

More and more manpower and resources are being devoted to the forces of nature—discovering, proving, understanding, using, and teaching these forces. But almost everyone agrees that one of the greatest forces on earth is *love*. Should churches finance constant research into this force of love? Should schools offer courses for credit (with homework, tests, and grades) on the force of love? The real wealth of a nation does not come from mineral resources but from the way it develops and harnesses the lovepower in the minds and hearts of its people.

In the New Testament an account of this force of love is presented very vividly and clearly:

Dear friends, let us love one another, because love is from God. Everyone who loves is a child of God and knows God; but the unloving know nothing of God. For God is love; and His love was disclosed to us in this, that He sent His only Son into the world to bring us life. The love I speak of is not our love for God, but the love He showed to us in sending His Son as the remedy for the defilement of our sins. If God thus loved us, dear friends, we in turn are bound to love one another. Though God has never been seen by any man, God Himself dwells in us if we love one another; His love is brought to perfection within us. (I John 4:7–12 NEB.)

God is love; he who dwells in love is dwelling in God, and God in him. This is for us the perfection of love, to have confidence on the day of Judgement, and this we can have, because even in this world we are as he is. There is no room for fear in love;

perfect love banishes fear. For fear brings with it the pains of judgement, and anyone who is afraid has not attained to love in its perfection. We love because He loved us first. But if a man says, "I love God," while hating his brother, he is a liar. If he does not love the brother whom he has seen, it cannot be that he loves God whom he has not seen. And indeed this command comes to us from Christ Himself; that he who loves God must also love his brother. (I John 4:12–21, NEB.)

If you love only those who love you, what credit is that to you? Even sinners love those who love them. Again, if you do good only to those who do good to you, what credit is that to you? Even sinners do as much. And if you lend only where you expect to be repaid, what credit is that to you: Even sinners lend to each other if they are to be repaid in full. But you must love your enemies and do good; and lend without expecting any return; and you will have a rich reward; you will be sons of the Most High, because He Himself is kind to the ungrateful and wicked. Be compassionate as your Father is compassionate.

Pass no judgement, and you will not be judged; do not condemn, and you will not be condemned; acquit, and you will be acquitted; give, and gifts will be given you. Good measure, pressed down, shaken together, and running over, will be poured into your lap; for whatever measure you deal out to others will be dealt to you in return. (St. Luke 6:32–38, NEB.)

And now I will show you the best way of all. I may speak in tongues of men or of angels, but if I am without love, I am a sounding gong or a clanging cymbal. I may have the gift of prophecy, and know every hidden truth; I may have faith strong enough to move mountains; but if I have no love, I am nothing. I may dole out all I possess, or even give my body to be burnt, but if I have no love, I am none the better.

Love is patient; love is kind and envies no one. Love is never boastful, nor conceited, nor rude; never selfish, not quick to take offence. Love keeps no score of wrongs; does not gloat over other men's sins, but delights in the truth. There is nothing love cannot face; there is no limit to its faith, its hope, and its endurance.

Love will never come to an end. Are there prophets? Their work will be over. Are there tongues of ecstasy? they will cease.

Is there knowledge? it will vanish away; for our knowledge and our prophecy alike are partial, and the partial vanishes when wholeness comes. When I was a child, my speech, my outlook and my thoughts were all childish. When I grew up, I had finished with childish things. Now we see only puzzling reflections in a mirror, but then we shall see face to face. My knowledge now is partial; then it will be whole, like God's knowledge of me. In a word, there are three things that last for ever: faith, hope and love; but the greatest of them all is love. (I Corinthians 12:31 to 13:13, NEB.)

As can be seen in these few passages from the Christian scriptures, many psycho-spiritual truths might be discovered by researching these age-old maxims about love. Maybe we will discover that love is indeed the basic force in the spiritual world. Maybe school children can be taught some laws of the spirit rather than slowly learning them later through suffering.

Are our prayers answered? On earth we will never know the reason why. Maybe it is an evidence of God's unlimited and undeserved love? God's love is its own ultimate reason. What is our response? Should it not be an overwhelming feeling of gratitude and desire to grow in some small way more like Him—to seek to be an open channel loving God and radiating God's love to all His children?

Love of God comes first and makes it easier to love in other ways. If we want our enemy to see only our good qualities and not our flaws, then lovingly look for and see only his good qualities. He, too, is a child of God. God loves us both even though neither of us is yet perfect. Above all, we should not only radiate love but also help others to become alive with love.

Throughout our discussion of the humble approach, we have been calling for and encouraging the expression of the highest and noblest qualities of the human spirit. Have we been asking too much? Is this plea for the humble approach too unrealistic to be achieved? We don't think so. It has always been our firm belief that human potential is far greater than most of us realize.

What are we—no more than creatures constituted by the possession of bodies and minds? Bodies and minds are only our tools. To the question *who are we* in ourselves, the answer should be given that the real self is a *soul*. Most people go to school for twelve long years just to educate the mind which lives only briefly. Can we not discover equally effective methods to educate the soul for eternity?

Two thousand years ago more time was devoted to spiritual education than to mental education. The same was true two hundred years ago. Most old universities were founded by ministers, to train ministers. But in the last two centuries we have increased mental education enormously, so that now twenty times as many hours are spent on mental as on spiritual improvement. No wonder the world is out of joint. Let us not work less for mental education; but could we not expect both children and adults to study at least seven hours a week for spiritual growth? The results might be rewarding.

It is not surprising that our world has problems. Many churches are no longer "relevant" because so little planning is devoted to spiritual education and so little time is actually set aside for it. If we tried to teach chemistry by such methods, there would be very few able chemists and few new discoveries in chemistry. Should not schools include in the curriculum courses in ethics, philanthropy, character-building, self-denial techniques, freedom from envy, joy of giving, thought-control, philosophy of life, etc.? To say that religion should not be included in university studies because it cannot be seen or accurately measured seems as questionable as saying that love should not be studied for the same reason.

New teaching methods could be tried. Maybe besides using only books and traditional classrooms we could teach effectively through newspapers, comics, radio, television, etc. At least church teachers could try to identify and warn against movies, magazines, and plays which are degenerate. Probably the new techniques of "programed learning" and "programed textbooks" could be adapted to teach spiritual growth at various ages, both in daily schools and Sunday schools.

Animals rely on instinct, so their mental development is slow. Many homes and many schools do not go much beyond instinct in their methods for teaching spiritual growth. For rational humans, education ought to be far more than merely a drawing out of what is somehow already embedded in the child as if by instinct.

By learning humility, we find that the purpose of life on earth is vastly deeper than any human mind can grasp. Diligently, each child of God should seek to find and obey God's purpose, but none be so egotistical as to think that he or she comprehends the infinite mind of God.

As we become more and more humble, we can learn more about God. Let us recall once more that man is able to observe only a tiny part of reality and his observations are often misleading because he is self-centered.

Scientists have steadily been changing their concepts of the universe and laws of nature, but the progression is always away from smaller self-centered or man-centered concepts. Evidence is always accumulating that things seen are only one aspect of the vastly greater unseen realities. Man's observational abilities are very limited, and so are his mental abilities. Should we not focus our lives on the unseen realities and not on the fleeting appearances? Should we not kneel down in humility and worship the awesome, infinite, omniscient, eternal Creator?

Every person's concept of God is too small. Through humility we can begin to get into true perspective the infinity of God. This is the humble approach. Are we ready to begin the formulation of a humble theology which can never become obsolete? This would be a theology really centered upon God and not upon our own little selves.

XIV   A New Research Program

I Humility Theology

Seeking to implement some of the ideas we have been considering in this book, we have begun to elaborate this idea of a humble theology. As we have said, it would be centered upon the infinitely creative God and so not be limited by our finite minds nor by past conceptions of what God should be like. "Humility Theology", as we now call it, has a special definition. Humility as people use the term generally is a fine virtue, but that is not what is intended here. We don't mean that you should be humble compared to other people, but rather that you should be humble about your knowledge of God.

Much of our knowledge of ultimate reality in this century seems to have come to us by way of scientific discovery. There is a great need for both scientists and theologians to be open to the spiritual significance of these new discoveries in the sciences. The vast and intricate universe which unfolds before us should challenge both theology and science to seek new avenues of interpretation and new areas of cooperation. Then, too, the emphasis upon humility is predicated upon the conviction that a change of heart is an essential element in this new thinking about religion. Much of our culture has become contentious, self-serving—even litigious—and that attitude has affected scientists and theologians and other thinking people in a way which makes honest exploration and open inquiry about God very difficult. There needs to be determined effort to bring our society to a personal confrontation with its own limitations and the deeper meaning of the universe and humanity's place in it.

Most theologies seem to have constructed a narrow framework within which God is allowed to act. Yet there is strong data to support the conclusion that we humans are part of a much, much larger framework for God's activity. There is also an enormous diversity in Nature, and we need to better appreciate that God may have far more interest in all of nature than we had supposed. Certainly our mutual interdependence with the rest of nature is becoming more apparent. And too, the enormity and diversity of the universe, and our ignorance about most of it, should help us to broaden our viewpoint about God's nature, plans and purposes.

To aid in this new search for truth, the John Templeton Foundation Inc of America and the Templeton Foundations in other nations have expanded in scope with the formation of a center for the study of humility theology with a view to promoting progress in religious thinking. The center has been named the Humility Theology Information Center, and has as two of its major components a series of research programs and an advisory board of prominent scientists and theologians interested in progress in religious thinking. A list of the Advisory Board is included in the Appendix to this book.

The research program which the Center has undertaken has three areas of concentration.

1. Utilization of scientific methods in understanding the work and purpose of the Creator.

2. Research on studying or stimulating progress in religion.

3. Research on the benefits of religion.

Examples of projects initiated by the Center include—

a. A bibliographic survey of work by scientists on spiritual subjects.

b. A program to assess the extent of teaching of university and college courses on science and religion and to stimulate courses emphasizing progress in religion.

c. A training module on religion and psychiatry which illustrates the extent to which spiritual factors may influence clinical therapy.

d. A program to encourage scientists and theologians to publish papers on humility theology.

e. A program of lectures on the relationship between science and theology presented at universities and colleges in North America and Europe and, more recently, at large churches in the United States.

At this early state of development, the Foundation's resources are directed principally to operating their own program initiatives, though they are developing mechanisms for advising others in finding support for their programs.

The general approach in carrying out the research projects of the Center has been to contract out the management of our programs to individuals with expertise in the area of each program. This approach will be illustrated briefly for several programs:

The National Institute for Healthcare Research, directed by Dr. David Larson, has carried out several studies for the Center researching the benefits of spirituality on health and well-being. The major source of research data has come from the clinical literature by a process called systematic review. The great value of this approach lies in its more rigorous methodology, applying strict, quantitative research methods which lead to objective results. This has been especially important in the case of social policy literature reviews, which previously have been done in a less systematic manner with frequently biased results.

Dr. Larson's published research data from systematic review, especially in the fields of psychiatry and primary care, has revealed sometimes a strong association between spiritual or religious commitment and imporved health. Given the national concerns for health care reform, these findings reveal that spirituality will have benefits not only in the general well-being of the individual but could also be a major factor in health care reform. Social policy issues are especially important here. For example, Dr. Larson's research group has examined the data which was obtained for the Surgeon General's study of the smoking habits of U.S. males, which clearly established the high risk for lung cancer and cardiovascular disease of two-pack-a-day smokers. However, Larson's group found that divorced individuals in the non-smoking group in the same study had the *same* cardiovascular risk as the heavy smokers. Divorce, so common and acceptable in our society, may be dangerous to an individual's health.

A second area which we have pursued is the preparation of a number of books describing research by scientists which may reveal the work and purposes of the Creator. In 1989, with Dr. Robert Herrmann, molecular biologist, as co-author, we published *The God Who Would Be Known,* a book dedicated to the pioneering research scientists who are discovering the accelerating creative processes. We now understand our universe and its life forms in ways which utterly challenge the capacity of our minds to comprehend. Yet at the same time we sense that maybe the universe has been made in such a way that its meaning and purpose somehow involve us and demand our attention and worship and even our participation. It is a book of wonder which can invoke in us a new kind of humility, a deep devotion to the Creator of the vast unseen. Then, in this year 1994, Dr. Herrmann and I have written a second book entitled *Is God the Only Reality?* which follows up on the ideas of the first book by examining the nature of scientific knowledge, examining aspects of research in modern physics and cosmology and the new understanding of evolution in terms of the self-organization of the cosmos. It seems that in each research area the mysteries are multiplied, as though what we measure with our instruments only hints at a vastly deeper network of meaning.

During this same period we have also edited a book by prominent scientists which is entitled *Evidence of Purpose.* Included are chapters by Paul Davies, John Polkinghorne, Arthur Peacocke, Owen Gingerich and Nobel laureate Sir John Eccles. Each has presented some facet of current scientific research which points beyond science to a plan or purpose in the universe.

It is hoped that these and future books on science's amazing discoveries will serve to encourage many more scientists and theologians and others to be attracted to the humble approach. We have also begun a series of programs to encourage educators and other scholars to direct their attention and the thinking of their students to the impact of science on theology.

Two years ago we began a program of prizes for scholarly papers which illustrate humility theology. At the same time we published a directory of scholars working on the relationship between science and theology. The first edition of this book, which is entitled *Who's Who in Theology and Science,* contains over 800 entries for individuals and some 60 organizations working in this field. The book is now under revision and the new edition will contain some 1200 individuals and over 75 organizations. The program of prizes for papers in humility theology—hereafter referred to as the Call for Papers program—has used this directory and several scholarly journal subscription lists to announce the program, and in the first two years over 100 papers and book chapters per year have been submitted for review. Prizes have been awarded in the same period for 66 scholarly papers which have been judged to demonstrate humility theology.

More recently we have begun a second program to encourage educators to emphasize the impact of science on religion in their teaching. An announcement was made in several journals and newsletters in the spring of 1993 which called for the submission of model courses in religion and science. Our request also carried the news that the 5 best courses would receive a cash prize and would be used as models for a larger program to encourage university, college and seminary teachers around the world to create new courses in science and religion along the lines of one or more of the model courses. We plan to award cash prizes for up to 100 such courses in 1995, and to couple the awards to a series of teaching workshops in the summer of that year.

Last year we also began another publication of the Humility Theology Information Center, a newsletter entitled *Progress in Theology*. Thus far it has functioned as the vehicle to carry news of the Center's Advisory Board and reports of events of immediate concern to the work of the Foundation. *Progress in Theology* has also published abstracts of lectures sponsored by The Foundation and abstracts of winning papers in the Call for Papers Program. There are presently about 1700 subscribers.

Finally in the past four years the Foundations have sponsored lectures on science and religion, with an emphasis on humility theology, in North America, Europe, and Australia and New Zealand.

Major lecture series have been sponsored by two U. S. organizations, the American Scientific Affiliation and the Chicago Center for Religion and Science. Lecture series have also been organized in the United Kingdom and in Eastern Europe by the British organization, Christians in Science, and another series has been presented in London under the auspices of The Royal Society for Arts, Manufactures and Commerce in cooperation with another British-based organization, The Science and Religion Forum. Finally, lectures have recently been presented in Germany by yet another science-religion organization, The European Society for the Study of Science and Theology.

II Progress in Religion

Though all of the above programs have been designed to elicit a new and progressive interest in religion, the Foundations and the Humility Theology Information Center continue to explore new areas of opportunity to extend religious thinking. Some of this new research presently has as its focus the development of a body of spiritual truth which is relatively free of sectarian or controversial content, and which is accepted worldwide regardless of the culture or the established religions of any geographical or ethnic area. There are two aspects to this newest initiative of the Center.

A. Discovering the Laws of Life

We have recently published a book entitled *Discovering the Laws of Life* which is a compilation of wise and practical sayings, laws which have a lasting value and a general acceptance among all people. It is planned to develop a variety of programs which may verify some of these laws scientifically—to provide them with solid scientific data as to their validity. Laws of this kind are truly basic and require no police, courts or judges. They're self-enforcing laws, in something like the way the law of gravity is self-enforcing. If you violate the law of gravity, it'll show you right away that you've violated the law. Now we don't mean by that that any one is going to discover laws that are a hundred percent effective immediately like the law of gravity, but we may discover laws that can be verified by statistical methods. Remember that fifty years ago most of us thought that smoking was harmful, but it's only in recent years that enough scientific studies have been done to demonstrate that you are more likely to die early if you smoke. Well, that is the sort of verification that we hope to achieve with some of these spiritual laws. We don't visualize that this body of laws will be permanent or complete any more than the laws of physics or genetics or anything else is permanent or complete. There should always be constant improvement, so we will have a body of laws of life that is always subject to criticism, verification and falsification. Some laws will be dropped because they were not as important or because they cannot be verified, while others will be added. By this means we may have an increasingly acceptable list of these laws of life.

The bottom line is that we are trying to have something in the field of spirituality that is not controversial from a sectarian standpoint. We're asking what you get out of religion, and one thing you get is information about how to lead a useful and happy life. Here then is a basic group of laws extracted from all religions and subject to verification which can be learned by people in all nations and which can be taught in a secular setting.

At present the laws of life are being tested in essay contests in grade school and high school populations in several parts of the United States and they are also being proposed for use in the British school system. We hope to see school courses develop—with homework, examinations, grades and credit toward graduation—around the world and especially in nations which now mandate the teaching of religion. We also anticipate a growing research program to seek to verify some of these laws of life.

B. Scientific Research on the Nature and Purposes of the Creator

The foregoing description of the laws of life leads logically to a series of questions about spiritual values possibly derived from science. For example, can we admire every ancient scripture and expand the benefits from ancient scripture while also building separately a science about purpose and creativity of which human purposes and creativity may be only one new manifestation? Could this yield a science respected worldwide and thereby free of the conflicts of ancient dogma? A theology for those who have no faith? Could this yield accelerating progress? Could we see universities include a course on purpose or creativity revealed by discoveries in science, like physics, astronomy, biology, archeology or medicine? Could courses include sections on expansion of concepts in space, duration, complexity, variety and on appearances vs. realities? Do the creations we study reveal aspects or purposes of a Creator? Maybe a new theology can develop from science—independent of ancient scripture and independent of ancient ecclesiology. It is said that God reveals himself both in the books of Scripture and also in the book of Nature. For centuries, enormous research has been done on the Scriptures. Only now are scientists beginning to say they may learn from their sciences more about the universal Creator, his purposes, and more about the universal spiritual laws that govern progress toward civilization. We have tremendous admiration for the enormous amount of work that's been done on ancient scripture. I served on the Board of Managers of the American Bible Society for fifteen years and have unlimited admiration for the ancient scriptures. Scriptures have been very beneficial to the whole world, but I am hoping we can develop a body of knowledge about God that doesn't rely on ancient revelations or scripture. It doesn't deny ancient scripture, it doesn't conflict with ancient scripture, but it doesn't rely on ancient scripture either. If there hadn't been any ancient scripture, what can we learn from what has been discovered in science?

Consider, for example, the size of God. At one time it would not have been imagined that God was a god of other planets besides the Earth. It's only in recent years that we've come to believe that there are a hundred billion other star systems like the sun in our own galaxy. And also that there are at least a hundred billion other galaxies. That's information about the creator, if we call the creator God or whatever we may choose to call the Creator. It's clear that the Creator is enormously larger than anybody had thought before, and that's just in a geographical sense. The other areas in which we discover knowledge about the Creator are numerous, including information about spiritual subjects such as love. Nobody denies that love is real, but how little we know about it. We know much more about your body than we know about the love you radiate, or the love you receive. Very little has been studied about the origin of love, the nature of love, the effects of love, the varieties of love. All those things could be studied scientifically and might lead to great improvements. The same thing might be done with prayer, forgiveness and gratitude. It could be done with many other spiritual subjects. We are trying to encourage development of a growing body of theology that is not controversial and not limited to any one area or any one ancient scripture, a theology respected by even the most skeptical scientist. A theology that is always progressing. In this way, we would hope to see spirituality made acceptable and respected universally.

Some of you may have read in the newspaper just recently that there has been a recommendation to the trustees of one of the great Lvy League schools, that they abolish the department of religion. The reason appears to be that most of the faculty do not think it's a respectable thing to teach religion at a university; that it's not scientific, or intellectual enough. Of course, we think that's a mistake. There are so many people who tend to think of religion as being unscientific that some of the great intellectual centers just neglect it. In fact, in Europe I'd say that the majority of people have the attitude that religion is not intellectually respectable. Now, because of that, we need to develop knowledge of God, an understanding of God, or an understanding of spiritual matters that is scientific, that is highly intellectual, and is not disputed because of divisions between religions or churches or ancient scripture or liturgy. In America alone, there are some six hundred departments of religion among universities and colleges. Most of these departments mainly study comparative religion or the history of religion similar to anthropology or archeology. They are backward looking.

If they were focused instead on new scientific evidence, or on purpose in life or on a purpose in the universe, or focused on aspects of God learned from the book of nature in addition to the book of scriptures, these departments of religion might be the most exciting part of the college campus. They might be constantly developing new and exciting concepts, new and exciting aspects of verification and falsification, so that the departments of religon would be a more influential part of college education. We want to help build a concept of teaching religion that is not localized. Perhaps without benefit of the wonderful ancient scriptures or religious organizations, we could encourage a new field of science seeking to learn aspects of the Creator by studying and proposing concepts to be verified or falsified such as the following: 1) Evidence of purpose. 2) Why is nature partly comprehensible? We do begin to understand some of the laws of nature, some of the mysteries of nature that are comprehensible to human beings. That itself is a mystery. Why is it comprehensible? 3) Why are humans creative? Why should humans have a desire to be creative? Why are humans dominated by purpose? All of these things we think should be studied so that in this generation many more discoveries may be made about the Creator, or God.

The main purpose of the Templeton Foundations is to encourage enthusiasm for accelerating discovery and progress in spiritual matters and in knowledge about the unlimited creative spirit. In conclusion, we don't think we've really discovered what to do yet. All of this is in the embryonic stage, in the formative stage, maybe about the same stage as many sciences were two centuries ago. That's very important to remember in the humble approach.

REFERENCES (FOOTNOTES)

1. *The Key to the Universe,* by Nigel Calder, 1977.

2. *In the Center of Immensities,* by Sir Bernard Lovell, 1978.

3. *Sartor Resartus,* by Thomas Carlyle, 1858.

4. *The Universe and Dr. Einstein,* by Lincoln Barnett, 1957.

5. *Natural Law in the Spiritual World,* by Henry Drummond, 1883.

6. *In Tune with the Infinite,* by Ralph Waldo Trine, 1897.

7. *Physics and Reality,* by Albert Einstein, 1936.

8. *An Introduction to Teilhard de Chardin,* by N. M. Wildiers, 1963.

9. *In the Center of Immensities,* by Sir Bernard Lovell, 1978.

10. *The Dragons of Eden,* by Carl Sagan, 1977.

11. "Science Pauses," by Vannevar Bush, *Fortune Magazine,* May, 1965.

12. *In Tune with the Infinite,* by Ralph Waldo Trine, 1897.

13. *The Transient and the Permanent in Christianity,* by Theodore Parker, 1841.

14. *Christianity in World History,* by Arend van Leewin, 1964.

15. *Empiricism: Scientific and Religious,* by Huston Smith, 1964.

16. *Something Beautiful for God,* by Malcolm Muggeridge, 1971.

17. *Spirits in Rebellion,* by Charles S. Braden, 1963.

18. "Science Pauses," by Vannevar Bush, *Fortune Magazine,* May, 1965.

19. "The Downfall of Communism," by Marceline Bradford, *The Freeman,* May, 1962.

20. *What is the Future of Man?* by Harold K. Schilling, 1971.

21. *New York Times,* March 5, 1970.

22. *Science and Human Values,* edited by Ralph Wendell Burhoe, 1971.

23. *The Dragons of Eden,* by Carl Sagan, 1977.

24. *Natural Law in the Spiritual World,* by Henry Drummond, 1883.

25. *Truth in Science and Religion,* by Henry Margenau, 1960.

26. *An Introduction to Teilhard de Chardin,* by Norbert M. Wildiers, 1968.

27. *Masterpieces of Christian Literature,* by Frank N. Magill, 1963.

28. *A Sermon at Unity Church, Miami,* by Charles Neal, 1976.

29. *My Philosophy and My Religion,* by Ralph Waldo Trine, 1896.

30. "The Church's One Foundation," by Samuel J. Stone, *Hymnal,* 1949.

31. *Sermon at Unity Church, Miami,* by Charles Neal, 1978.

32. *What All the World's A'seeking,* by Ralph Waldo Trine, 1896.

33. *The Divine Imperative,* by Emil Brunner, 1932.

34. *Architects of Moral Anarchy,* by John Howard, 1978.

KJV = King James Version

RSV = Revised Standard Version

NEB = New English Bible

JB   = Jerusalem Bible.

Bibliography

Barbour, I. G.
Christianity and the Scientist
New York: Association Press, 1960
Issues in Science and Religion
Englewood Cliffs, N. J.: Prentice-Hall, 1965
Myths, Models and Paradigms: A Comparative Study in Science and Religion
New York: Harper and Row, 1974
Science and Secularity: The Ethics of Technology
New York: Harper and Row, 1970

H. Benson
The Relaxation Response
New York: William Morrow, 1975

David Bohm
Wholeness and the Implicate Order
London: Routledge and Kegan Paul, 1980

Briggs, John P., and F. David Peat
Looking Glass Universe: The Emerging Science of Wholeness
New York: Simon and Schuster, 1984

Bube, R. H.
The Human Quest: A New Look at Science and the Christian Faith
Waco, Texas: Word, 1971
The Encounter between Christianity and Science
Grand Rapids: Eerdmans, 1967

Burnoe, Ralph Wendell (ed.)
Science and Human Values in the 21st Century
Philadelphia: Westminster Press, 1971

Cantore, Enrico
Scientific Man: The Humanistic Significance of Science
New York: ISH Publications, 1977

Capra, Fritjof
The Tao of Physics: An Exploration of the Parallels between Modern Physics and Eastern Mysticism
Berkeley: Shambhala, 1977

Clark, Robert Edward David
The Universe and God: A Study of the Order of Nature in the Light of Modern Knowledge
London: Hodder and Stoughton, 1939
Scientific Rationalism and Christian Faith, with particular reference to the writings of Professor J. B. S. Haldane and Dr. J. S. Huxley
London: Inter-Varsity Fellowship, 1945
The Universe: Plan or Accident? The Religious Implications of Modern Science
London: Paternoster Press, 1949

Coulson, Charles Alfred
Christianity in an Age of Science
London: G. Cumberlege, Oxford, 1953
Science and Christian Belief
London: Oxford University Press, 1955

Cupitt, Don
The Worlds of Science and Religion
New York: Hawthorn Books, 1976; London: Sheldon Press

Davies, Paul
God and the New Physics
New York: Simon & Schuster, 1983
The Cosmic Blueprint
New York: Simon & Schuster, 1988

Dobzhansky, Theodosius Grigorievich
The Biology of Ultimate Concern
New York: World, 1971
"Evolution as a Creative Process"
Proceedings of the Ninth International Congress of Genetics
Caryologie Suppl., 1954– pp. 435–48

Dolphin, Lambert
Lord of Time and Space
Westchester, Illinois: Good News Pub., 1974

Dyson, Freeman
Infinite in All Directions
New York: Harper and Row, 1988

Eccles, J. C.
Facing Reality: Philosophical Adventures of a Brain Scientist
Berlin, Heidelberg, and New York: Springer-Verlag, 1970
The Human Mystery
The Gifford Lectures, University of Edinburgh, 1979
The Wonder of Being Human
Boston: New Science Library, Shambhala Publications, 1985

Eccles, J. C., with Karl R. Popper
The Self and Its Brain: An Argument for Interactionism
Berlin, Heidelberg, and New York: Springer International, 1977

Eddington, Arthur Stanley
The Nature of the Physical World
Cambridge: University Press, 1928
Science and the Unseen
New York: Macmillan, 1930

Eiseley, Loren
The Firmament of Time
New York: Atheneum, 1984
The Invisible Pyramid
New York: Charles Scribner and Sons, 1970
The Immense Journey
New York: Vintage Books, Random House, 1957

Ferris, Timothy
Coming of Age in the Milky Way
New York: William Morrow, 1988

Gilkey, Langdon
Maker of Heaven and Earth
Garden City, N. Y.: Doubleday, 1959
Religion and the Scientific Future: Reflections on Myth, Science and Theology
New York: Harper and Row, 1970

Gould, S.
Wonderful Life: The Burgess Shale and the Nature of History
New York: W. W. Norton & Co., 1989

Haldane, John Burdon Sanderson
Possible Worlds and Other Papers
Freeport, N. Y.: Books for Libraries Press, 1971 (1928)

Hardy, Alister Clavering
The Living Stream
London: Collins, 1965
The Divine Flame: An Essay Towards a Natural History of Religion
London: Collins, 1966

The Biology of God: A Scientist's Study of Man and the Religious Animal
London: J. Cape, 1975
The Spiritual Nature of Man
Oxford University Press, 1979

Hartshorne, Charles
A Natural Theology for our Time
La Salle, Illinois: Open Court, 1967
The Divine Relativity: A Social Conception of God
New Haven: Yale University Press, 1948

Hawking, Stephen
A Brief History of Time
Toronto: Bantam Books, 1988

Hefner, Philip
*The Promise of Teilhard: The Meaning of the Twentieth Century in
  Christian Perspective*
Philadelphia: J. B. Lippincott, 1970
"The Future as Our Future: A Teilhardian Perspective," in E. H.
  Cousins (ed.),
Hope and the Future of Man
Philadelphia, Fortress Press, 1972
"Basic Assumptions about the Cosmos," in W. Yourgrau and A. D.
  Breck
Cosmology, History and Theology
New York: Plenum Press, 1977

Heim, Karl
Christian Faith and Nature Science
London: SCM Press, 1953
The Transformation of the Scientific World View
London: SCM Press, 1954

Heisenberg, Werner
Across the Frontiers
New York: Harper Torchbooks, 1971
Physics and Beyond
London: George Allen and Unwin, 1971
Physics and Philosophy: The Revolution in Modern Science
St. Andrews Gifford Lectures, 1955–56
New York: Harper, 1958

Hesse, Mary Brenda
*Science and the Human Imagination: Aspects of the History and Logic of
  Physical Science*
New York: Philosophical Library, 1955
Revolutions and Reconstructions in the Philosophy of Science
Brighton: Harvester Press, 1981

Hodgson, P. E.
Is Science Christian?
London: Catholic Truth Society, 1978

Hooykaas, Reijer
Philosophia Libera: Christian Faith and the Freedom of Science
Wheaton, Illinois: Tyndale Press, 1957
Religion and the Rise of Modern Science
Edinburgh: Scottish Academic Press, 1957
Natural Law and Divine Miracle: A Historical-Critical Study of the Principle of Uniformity in Geology, Biology and Theology
Leiden: E. J. Brill, 1959
The Christian Approach in Teaching Science
London, Published for the Research Scientists' Christian Fellowship by the Tyndale Press, 1960

Jaki, Stanley
The Origin of Science and the Science of Its Origin
Edinburgh: Scottish Academic Press, 1978
Planets and Planetarians
Edinburgh: Scottish Academic Press, 1978
The Road of Science and the Ways to God
Edinburgh: Scottish Academic Press, 1978

Kaiser, Christopher B.
The Logic of Complementarity in Science and Theology
Ph.D. dissertation, University of Edinburgh, 1974

Kass, J., R. Friedmann, J. Leserman, P. Zuttermeister, and H. Benson
"Health Outcomes and a New Index of Spiritual Experience,"
Journal for the Scientific Study of Religion, in press.
S. Myers and H. Benson, "Psychological Factors in Healing: A New Perspective on an Old Debate," *Behavioral Medicine* 18 (1992)

Khan, Mohammed Yamin
God, Soul, and Universe in Science and Islam
Lahore: Sh. M. Ashraf, 1962

Larson, D., E. Patterson, D. Blazer, A. Omran, and B. Kaplan
"Systematic Analysis of Research on Religious Variables in Four Major Psychiatric Journals, 1978–1982," *American Journal of Psychiatry* 143 (1986) 329–34.
F. Craigie, D. Larson, and I. Liu, "References to Religion in the Journal of Family Practice," *Journal of Family Practice* 30 (1990)

Lever, Jan
Where are we Headed? A Biologist Talks about Origins, Evolution, and the Future
(Tr. Walter Lagerway), Grand Rapids: Eerdmans, 1970

Lonergan, Bernard
Insight: A Study in Human Understanding
London: Longmans, Green, 1957
Method in Theology
London: Darton, Longman and Todd, 1971
Philosophy of God and Theology
London: Darton, Longman and Todd, 1974

Lovell, Bernard
The Individual and the Universe: The 1958 Reith Lectures
Oxford: Oxford University Press, 1959
The Exploration of Outer Space
London: Oxford University Press, 1962
In the Center of Immensities
New York: Harper and Row, 1978; London, Hutchinson, 1979

MacKay, Donald MacCrimmon (ed.)
Christianity in a Mechanistic Universe
Chicago: Inter-varsity Press, 1965

MacKay, Donald MacCrimmon
Information, Mechanism and Meaning
Cambridge, Mass.: M.I.T. Press, 1969
The Clock-work Image: A Christian Perspective on Science
Downers Grove, Illinois: Inter-varsity Press, 1974
Freedom of Action in a Mechanistic Universe
London: Cambridge University Press, 1976
Science, Chance and Providence: The Riddell Memorial Lectures delivered at the University of Newcastle upon Tyne, March 1977
New York: Oxford, 1978
Human Science and Human Dignity: London Lectures in Contemporary Christianity
London: Hodder and Stoughton, 1979
Brains, Machines and Persons
London: Collins, 1980; Grand Rapids: Eerdmans, 1980

Mascall, Eric Lionel
Christian Theology and Natural Science: Some Questions of Their Relations
London: Longmans, Green, 1956
The Secularisation of Christianity
London: Darton, Longman and Todd, 1965
The Openness of Being: Natural Theology Today
London: Darton, Longman and Todd, 1971

McMullin, Ernan
"A Case for Scientific Realism, in *Scientific Realism,* ed. Jarrett Leplin
Berkeley: University of California Press, 1984

Morris, Richard
The Nature of Reality
New York: Noonday Press, Farrar Straus and Giroux, 1987

Pannenberg, W.
The Idea of God and Human Freedom
Philadelphia: Westminster Press, 1973
Theology and the Philosophy of Science
Tr. Francis McDonagh, Philadelphia: Westminster press, 1976

Peacocke, Arthur Robert
Science and the Christian Experiment
London: New York, Oxford, 1971
Creation and the World of Science
Oxford: Oxford University Press, 1979
"Cosmos and Creation, in *Cosmology, History and Theology*, pp. 365–81
    ed. W. Yourgrau and A. D. Breck, New York: Plenum Press, 1977
"The Nature and Evolution of Biological Hierarchies," in *New Approaches to Genetics*, pp. 245–304, ed. P. W. Kent
London: Oriel Press, Routledge and Kegan Paul, 1978
Intimations of Reality: Critical Realism in Science and Religion
Notre Dame, Ind.: University of Notre Dame Press, 1984
God and the New Biology
San Francisco: Harper and Row, 1986

Polanyi, Michael
Science, Faith and Society
London: Oxford University Press, 1946
The Logic of Liberty
London: Routledge and Kegan Paul, 1951
Personal Knowledge: Towards a Post-Critical Philosophy
London: Routledge and Kegan Paul, 1958
The Study of Man
London: Routledge and Kegan Paul, 1959
The Tacit Dimension
London: Routledge and Kegan Paul, 1967
Knowing and Being
London: Routledge and Kegan Paul, 1969
Scientific Thought and Social Reality: Essays by Michael Polanyi, edited by
    Fred Schwarz, Psychological Issues Volume VIII/No. 4, Monograph 32
New York: International Universities Press, 1974

Polanyi, Michael, with Harry Prosch
Meaning
Chicago: Chicago University Press, 1975

Polkinghorne, John
The Way the World Is
Grand Rapids: Eerdmans, 1986
The Quantum World
London: Longmans, 1984
Science and Creation: The Search for Understanding
Boston: New Science Library, 1989
Reason and Reality
London: SPCK, 1991

Pollard, William Grosvenor
The Cosmic Drama
New York: National Council of the Episcopal Church, 1955
Chance and Providence: God's Action in a World Governed by Scientific Law
New York: Scribner's, 1958
Physicist and Christian: A Dialogue between the Communities
Greenwich, Conn.: Seabury Press, 1961
Science and Faith: Twin Mysteries
New York: T. Nelson, 1970
"Rumors of Transcendence in Physics," *America Journal of Physics* 52
   (1984) 877–81

Prigogine, Ilya and Stengers, Isabelle
Order out of Chaos
New York: Bantam Books 1984

Ramm, Bernard
The Christian View of Science and Scripture
Grand Rapids: Eerdmans, 1954

Ramsey, Ian Thomas
Religion and Science, Conflict and Synthesis: Some Philosophical Reflections
London: S.P.C.K., 1964
Models and Mystery
Oxford: Oxford University Press, 1964
Christian Empiricism
London: Sheldon Press, 1974

Ramsey, Ian Thomas (ed.)
*Biology and Personality: Frontier Problems in Science, Philosophy and Re-
   ligion*
Oxford: Blackwell, 1965

Raven, Charles Earle
Evolution and the Christian Concept of God
London: Oxford University Press, 1936
Science and Religion
Cambridge: Cambridge University Press, 1953

Christianity and Science
New York: Association Press, 1955
Science, Medicine and Morals: A Survey and a Suggestion
London: Hodder and Stoughton, 1959

Roy, Rustum
Experimenting with Truth
Oxford: Pergamon Press, 1981

Ridderbos, Nicolaas Herman
Is there a Conflict between Genesis 1 and Natural Science?
Grand Rapids: Eerdmans, 1957

Rimmer, Harry
The Harmony of Science and Scripture
Grand Rapids: Eerdmans, 1949 (1936)

Schrodinger, Erwin
Mind and Matter
London: Cambridge University Press, 1958
What is Life?
New York: Doubleday/Anchor, 1956

Shapley, Harlow (ed.)
Science Ponders Religion
New York: Appleton-Century-Crofts, 1960

Soskice, Janet
Metaphor and Religious Language
Oxford: Clarendon Press, paperback edition, 1987

Teilhard de Chardin, Pierre
Activation of Energy
Tr. Rene Hague, London: Collins, 1965; New York: Harcourt Brace
    Jovanovich, 1971
Science and Christ
Tr. Rene Hague, London: Collins, 1968
Hymn of the Universe
Tr. Gerald Vann, London: Collins, 1965; New York: Harper and
    Row, 1969
The Divine Milieu
Gen. ed. Bernard Wall, New York: Harper and Row, 1965
The Phenomenon of Man
Tr. Bernard Wall, New York: Harper and Row, 1959

Templeton, John and Herrmann, Robert
The God Who Would Be Known
New York: Harper and Row, 1989
Is God the Only Reality?
Continnum, New York 1994

Torrance, Thomas F.
Belief in Science and in Christian Life
Edinburgh: Handsel Press, 1980
The Christian Frame of Mind
Edinburgh: Handsel Press, 1985
Divine and Contingent Order
Oxford: Oxford University Press, 1981
The Ground and Grammar of Theology
Charlottesville: University of Virginia Press, 1980
Theology in Reconstruction
London: SCM Press, 1965
Space, Time and Incarnation
London: Oxford University Press, 1969
Theological Science, Based on the Hewett Lectures for 1959
London: Oxford University Press, 1969
God and Rationality
London: Oxford University Press, 1971
Space, Time and Resurrection
Edinburgh: Handsel Press and Grand Rapids: Eerdmans, 1976

von Weizsacker, C. F.
The History of Nature
Chicago: University of Chicago Press, 1949
The Relevance of Science: Creation and Cosmogony
London: Collins, 1964

Weinberg, Steven
The First Three Minutes: A Modern View of the Origin of the Universe
London: Andre Deutsch, 1977

Whitehead, Alfred North
Process and Reality: An Essay in Cosmology
Cambridge: Cambridge University Press; New York, Macmillan, 1929
Adventures of Ideas
Cambridge: Cambridge University Press, 1933
Modes of Thought
Cambridge: Cambridge University Press, 1938

Young, Louise
The Unfinished Universe
New York: Simon & Schuster, 1986

Humility Theology
Information Center
A Limited Explanation

The John Templeton Foundation has established as one of its primary goals the promotion of a greater awareness among thinking people—especially scientists and theologians—of the vast magnitude of the Creator God and the enormity of our own ignorance. The attitude of humility which such an awareness demands might have profound impact upon the goals of both scientists and theologians. There could be fresh openness to spiritual dimensions—with scientists focusing more of their energies on research into the spiritual dimension, and with theologians contributing fresh interpretations and new directions of theological explanation.

The increasing appreciation of the vastness and intricacy of the universe and the subtlety of its interrelationships which science reveals carries with it profound theological implications. This new theological outlook is part of the theology of *humility*. For this purpose the Foundation has instituted a *Humility Theology Information Center*. The Center's current focus includes the following areas of concentration

- Utilization of Scientific Methods in Understanding the Work and Purpose of the Creator

- Research in Studying or Stimulating Progress in Religion

- Research in Benefits of Religion

The possibilities from addressing these and many other questions can be illuminating. And in all of this, the spirit of humility seems essential. This unlimited investigation and promotion is the major emphasis of the *Humility Theology Information Center*.

APPENDIX 2

Excerpt from
Riches for the Mind and Spirit
Edited by John Marks Templeton

We are perched on the frontiers of future knowledge. Even though we stand upon the enormous mountain of information collected over the last five centuries of scientific progress, we have only fleeting glimpses of the future. To a large extent, the future lies before us like a vast wilderness of unexplored reality. The God who created and sustains His evolving universe through eons of progress and development has not placed our generation at the tag end of the creative process. He has placed us at a new beginning. We are here for the future.

Our role is crucial. As human beings we are endowed with mind and spirit. We can think, imagine, and dream. We can search for future trends through the rich diversity of human thought. God permits us in some ways to be co-creators with Him in His continuing act of creation.

Scientists have steadily been changing their concepts of the universe and laws of nature, but the progression is always away from smaller self-centered or human-centered concepts. Evidence is always accumulating that things seen are only one aspect of the vastly greater unseen realities. Human observational abilities are very limited and so are our mental abilities. Should we not focus ourselves on the unseen realities and not on the fleeting appearances? Should we not kneel down in humility and worship the awesome, infinite, omniscient, eternal Creator?

By learning humility, we find that the purpose of life on earth is vastly deeper than any human mind can grasp. Diligently, each child of God should seek to find and obey God's purpose, but none be so egotistical as to think that he or she comprehends the infinite mind of God.

Every person's concept of God is too small. Through humility we can begin to get into true perspective the infinity of God. This is the humble approach. Are we ready to begin the formulation of a humble theology which can never become obsolete? This would be a theology really centered on God and not our own little selves.

John Marks Templeton

Board of Advisors of the John Templeton Foundation Humility Theology Information Center

Fall 1997

North America

Mrs. Elizabeth Peale Allen, Vice Chairman of the Peale Center for Christian Living in Pawling, NY. She is Chairman of the Board of the Positive Thinking Foundation.

Professor Francisco J. Ayala, Donald Bren Professor of Biological Sciences and Professor of Philosophy at the University of California, Irvine. He is a member of the President's Committee of Advisors on Science and Technology and has been President and Chairman of the Board of the American Association for the Advancement of Science and the Society for the Study of Evolution. His research focuses on population and evolutionary genetics, including the origin of species, genetic diversity of populations, and the molecular clock of evolution.

Dr. Ian Barbour is Professor Emeritus in Physics and Religion at Carleton College in Northfield, Minnesota. In addition to serving as a Gifford Lecturer in 1989-91, he has written several books addressing the interface of religion and science including *Religion in an Age of Science* and *Ethics in an Age of Technology.*

Dr. Herbert Benson, President of the Mind/Body Medical Institute and Associate Professor of Medicine at Harvard Medical School. He has studied the efficacy of religion and spirituality upon physical health and healing. He has written numerous books including *The Relaxation Response,* and *Beyond the Relaxation Response.*

Dr. Keith Briscoe, past President of Buena Vista College in Iowa in 1974-1994. In the years since he has worked tirelessly to lead Buena Vista College into the twenty-first century as one of the truly outstanding small colleges in America.

Dr. Freeman Dyson is Professor of Physics at the Institute for Advanced Study, Princeton, NJ. He has received numerous honors and been widely published for his work in physics and ethics in science with regard to arms control. Dr. Dyson is a fellow of the Royal Society of London and a 1985 Gifford Lecturer. He received the Phi Beta Kappa Award in Science in 1988 for his book, *Infinite in All Directions.*

Dr. Lindon Eaves, Distinguished Professor of Human Genetics and Professor of Psychiatry at the Medical College of Virginia. Dr. Eaves is also an ordained priest of the Episcopal Church. He has published extensive research involving genetic studies of twins and has also published on the interface between religion and science.

Dr. Diana L. Eck is Professor of Comparative Religion and Indian Studies at Harvard University. She is also Chair of the Committee on the Study of Religion in the Faculty of Arts and Sciences and a member of the Faculty of Divinity. Her extensive work on India includes the book *Banaras, City of Light.* In 1995, Dr. Eck was awarded the Henry Luce III Fellowship in Theology for her work on a book entitled *Multireligious America: New Challenges for American Pluralism.*

Mr. Foster Friess is Chairman of Friess Associates, managers of over $12 billion of equities. A graduate of the University of Wisconsin, Friess currently serves on the Advisory Council of the Royal Swedish Academy of Sciences of Stockholm, which awards the Nobel Prize for chemistry and physics, and the Executive Committee of the Council for National Policy, which networks leaders in the U.S. who are committed to a strong national defense, traditional values and the free enterprise system.

Professor Owen Gingerich is Professor of Astronomy and of the History of Science at the Harvard-Smithsonian Center for Astrophysics in Cambridge, Massachusetts. He is a member of the American Philosophical Society, the American Academy of Arts and Sciences, and the International Academy of the History of the Sciences. Professor Gingerich has published 500 technical or educational articles and reviews.

Mr. Kenneth Giniger, President of the K.S. Giniger Company, Inc., has co-published several books with Mr. Templeton that address science and religion. He is chairman of the Layman's National Bible Association.

Rev. Philip Hefner, Professor of Systematic Theology, Lutheran School of Theology at Chicago. He earned his Ph.D. with distinction from the University of Chicago and has taught at Lutheran seminaries his entire career. He currently serves as Co-Director of the Chicago Center for Religion and Science and is Editor-in-Chief of *Zygon: Journal of Religion* and Science. His most recent book is *The Human Factor: Evolution, Culture and Religion.*

Dr. Robert Herrmann is Adjunct Professor of Chemistry and Chair of the Premedical Program of Gordon College and past Executive Director of The American Scientific Affiliation. His work, both independent and through collaborative writing efforts with John M. Templeton, has sought to encourage greater spirituality among professionals in the natural and physical sciences. He is Editorial Coordinator for the *Progress in Theology* newsletter and Director of the Science-Religion Course Programs. He is a trustee of the John Templeton Foundation.

Dr. Harold G. Koenig is Director of the Program on Religion, Aging, and Health at the Center for Aging, Duke University Medical Center. He is also Assistant Professor of Medicine and Psychiatry, Division of Geriatric Medicine, Duke University Medical Center. In 1995 he became a Diplomat of the American Board of Psychiatry and Neurology.

Dr. David Larson, a psychiatrist, is a former Senior Fellow at the National Institute for Mental Health (NIMH). He is currently President of the National Institute for Healthcare Research. He has published journal articles and a Psychiatric Training Manual which demonstrate that spirituality and religious practice can benefit physical and mental health and healing.

Mr. Robert F. Lehman is President and CEO of the Fetzer Institute and Chairman of the Board of the Fetzer Memorial Trust. He has been Vice President and Director of International Programs at the Kettering Foundation and Director of the Exploratory Fund that sponsored research on the relationship of consciousness to health and education.

Dr. Martin Marty is a Senior Scholar at The Park Ridge Center for the Study of Health, Faith and Ethics in Chicago, IL. He is the editor of

Second Opinion, a journal providing a forum for the interface of health, faith, and ethics.

Mr. Gary D. Moore is a counselor to ethical and religious investors with twenty years of Wall Street experience, including service as Senior Vice President of Investments for Paine-Webber. Mr. Moore founded Gary Moore & Co., a company dedicated to providing counsel to ethical and religious investors. He currently counsels some of America's best-known churches and banks. He is also a commentator on the political economy for UPI National Radio. Mr. Moore authored *The Christian Guide to Wise Investing.*

Rev. Dr. Glenn Mosley is President and CEO of the Association of Unity Churches. His ministry began in 1957 and he has traveled extensively speaking in Unity and Non-Unity Churches. In 1964, he began a television ministry and has appeared on radio and/or television daily for 21 years. He frequently serves as a visiting professor for colleges and universities and conducts workshops on interpersonal communications and life and death transitions. He is a trustee of the John Templeton Foundation.

Dr. Nancey Murphy is Associate Professor of Christian Philosophy at Fuller Theological Seminary in Pasadena, CA. Her thought-provoking articles and books about religion in the age of science have been well received by critical reviewers. She has written a series on the philosophy of religion entitled *Theology in the Age of Scientific Reasoning.*

Dr. David G. Myers, John Dirk Werkman Professor of Psychology at Hope College, Holland, MI. Dr. Myers has published ten books and two best-selling psychology textbooks. He is the recipient of the Gordon Allport Prize for his National Science Foundation-funded experiments on group influence. His most recent publication is the popular *The Pursuit of Happiness: Who is Happy—and Why.* He is a trustee of the John Templeton Foundation.

Dr. Seyyed Hossein Nasr is University Professor of Islamic Studies at George Washington University in Washington, D.C. He was the first Muslim to give the Gifford Lectures in 1981. Dr. Nasr is the author of over twenty books and over two hundred articles. He has lectured extensively throughout the Islamic world, Western Europe, North and Central America, India, Japan and Australia and has participated in numerous conferences and congresses on Islam, philosophy, comparative religion and the environmental crisis.

Dr. Ravi Ravindra is Professor and Chair of Comparative Religion, Professor of International Development Studies and Adjunct

Professor of Physics at Dalhousie University, Halifax, Canada. He was Founding Director of the Threshold Award for integrative knowledge and chair of its international and inter-disciplinary Selection Committees in 1979 and 1980. His most recent book is entitled, *Krishnamurti: Two Birds in One Tree.*

Mr. Laurance S. Rockefeller, philanthropist, business executive and conservationist, has held various Chairs and Trustee appointments for a number of national, academic, and humanitarian organizations. Mr. Rockefeller has been the recipient of many awards and medals for his conservation work and philanthropic interests, including the Congressional Gold Medal in 1991.

Dr. Beverly Rubik is the Founder and Executive Director of the Institute for Frontier Science, at Temple University, Philadelphia. She earned her Ph.D. in biophysics from the University of California at Berkeley and has published two books, over 30 papers, and lectures nationally and internationally on complementary medicine and science and spirit. She serves on the editorial or advisory boards of 5 peer-reviewed journals on complementary medicine.

Dr. Robert John Russell is Founder and Director of the Center for Theology and the Natural Sciences (CTNS) and Professor of Theology and Science in residence at The Graduate Theological Union in Berkeley, California. He received his Ph.D. in physics from the University of California at Santa Cruz, and his specialized scientific interests include cosmology and thermodynamics. Dr. Russell has co-edited three books as a result of ongoing collaborative research between CTNS and the Vatican Observatory.

Dr. Kevin Sharpe is Core Professor, Graduate School, The Union Institute, and Founding Editor of *Science and Religion News,* published by the Institute on Religion in an Age of Science. He has written prolifically on the subject of science and religion and currently is Series Editor for *Theology and the Sciences* series for Fortress Press.

Dr. Howard Van Till is Professor and Chairman of the Department of Physics at Calvin College, Michigan. He is a member of the American Astronomical Society and the American Scientific Affiliation. He has written books and articles addressing creation and cosmology from a Christian perspective.

Dr. Everett L. Worthington is a professor of Psychology at Virginia Commonwealth University. In 1986 he published an extensive review of research on religion in counseling and recently provided a ten-year update to this review, which was published in *Psychological Bulletin*. Dr. Worthington has published upwards of 75 scientific articles in peer-reviewed journals on the role of spirituality in health and health-care.

International

Dr. M A Zaki Badawi is Principal of the Muslim College in London. He is Chairman of the UK Imams and Mosques Council and the UK Muslim Law (Shariah) Council. He is a lecturer at Al-Azhar University, Cairo. Dr. Badawi frequently writes and broadcasts on Muslim affairs.

Professor John D. Barrow is Professor of Astronomy in the Physics and Astronomy Department at the University of Sussex and is considered a leading writer of book and articles on efforts to understand the universe. He is the author of *Pie in the Universe, The Anthropic Cosmological Principal* (with Frank Tipler), *The World Within the World,* and *The Artful Universe.*

Dr. Paul Davies is a mathematical physicist and author of over 20 books. He has written extensively on God's relationship to science and the universe, notably *The Cosmic Blueprint* and *The Mind of God.* He is a professor of theoretical physics at Adelaide University in Australia and the 1995 recipient of the Templeton Prize for Progress in Religion.

Professor George F. R. Ellis has been a visiting lecturer and professor in cosmology, physics and astronomy across the globe, including South Africa, England, Germany, Canada, Italy and the United States. He received his Ph.D. in Applied Maths and Theoretical Physics from the St. John's College at Cambridge, is a fellow of the Royal Society of South Africa, and was President of the International Society of Relativity and Gravitation. Dr. Ellis is co-author of *On the Moral Nature of the Universe: Theology, Cosmology, and Ethics (Theology and the Sciences)* with fellow Foundation Advisor Nancey Murphy, and has collaborated with Stephen Hawking on a number of publications including *The Large Scale Structure of Space-Time.*

The Rt. Hon. John Selwyn Gummer is an outstanding Christian Layman, member of the House of Commons, and past Chairman of the Conservative Party. Mr. Gummer has been chief executive in several publishing companies and a leader in publishers associations. He has written several books. Mr. Gummer has held several British

Cabinet portfolios and is a member of the General Synod of The Church of England.

Professor Malcolm Jeeves, Research Professor of Psychology, University of St. Andrews, is President of The Royal Society of Edinburgh and was made Commander of the Order of the British Empire in the Queen's National New Years Honours in 1992 for his services to science and to psychology in Britain. He established the Department of Psychology at St. Andrews University and his research interests center around cognitive psychology and neuropsychology.

Rev. Dr. Stephen Orchard is the Director of Christian Education Movement and a minister of the United Reformed Church. He is the editor of *REToday*. Rev. Orchard was Assistant General Secretary of the British Council of Churches and is a Director and Vice-Chairman of the United Kingdom Temperance Alliance Ltd.

Rev. Dr. Sir John C. Polkinghorne F.R.S. President of Queen's College, Cambridge is a member of the Church of England Doctrine Committee and General Synod. Former Professor of Applied Physics at Cambridge, he has published many papers on theoretical elementary particle physics. Among his science and religion books are *Science and Creation* and *Reason and Reality*.

Professor F. Russell Stannard is currently the Vice President of the Institute of Physics and Professor and Department Head of Physics at the Open University, UK. He has served on the Prime Minister's Advisory Committee on Science and Technology in the UK. Professor Stannard has authored the popular "Uncle Albert" series introducing physics to young children. He is the author of *Grounds for Reasonable Belief,* the twelfth book in the series, *Theology and Science.* He is a trustee of the John Templeton Foundation.

Professor Keith Ward is Regius Professor of Divinity at the University of Oxford and formerly Professor of History and Philosophy of Religion at King's College, London University. He is one of the country's foremost writers on comparative religion and Christian issues. A Gifford Lecturer, whose most recent book, *Defending the Soul,* is an affirmation of human divinity and value.

Members

Mr. Baba Amte
Prof. Charles Birch
Rev. Canon Michael Bourdeaux
Mrs. Wendy Brooks
Mrs. Amy Van Hoose Butler
Mrs. Ann Cameron
Mr. Douglas Cameron
Mr. Robert Handly Cameron
Mrs. Leigh Cameron
Miss Jennifer Cameron
Dr. Paul Davies
Mr. Charles R. Fillmore
Mrs. Connie Fillmore Bazzy
Ms. Wilhelmina Griffiths
Dr. Kyung-Chik Han
Mr. R.M. Hartwell
Dr. Robert Herrmann
Prof. Colin Humphreys
Prof. Stanley Jaki
Rt. Hon The Lord Jakobovits
Dr. Inamulla Khan
Rev. Dr. Bryant Kirkland
Mrs. Avery Lloyd
Mr. David W. Lloyd
Miss Chiara Lubich
Rev. Dr. Glenn Mosley
Dr. Donald Munro
Rev. Nikkyo Niwano
Mr. Michael Novak
Prof. Colin Russell
Prof. Pascal Salin
Dame Cicely Saunders

Mrs. Frances Schapperle
Brother Roger Schutz-Marsauche
Miss Lise Shapiro
Mrs. Jill Sideman
Mr. Richard Sideman
Ms. Renee Dawn Stirling
Mr. Christopher Templeton
Mr. Handly Templeton
Mrs. Rebecca L. Templeton
Mrs. Rebecca M. Templeton
Miss Lauren Templeton
Mr. Harvey M. Templeton, Jr.
Mr. Harvey M. Templeton, III
Miss Heather Templeton
Miss Jennifer A. Templeton
Mrs. Jewel Templeton
Sir John M. Templeton
Dr. John M. Templeton, Jr.
Dr. Josephine Templeton
Rev. Prof. Thomas Torrance
Miss Elizabeth Transou
Mr. Stamps Transou
Prof. Dr. Carl Friedrich von Weizsäcker
Ms. Mary P. Walker
Mr. Linford G. Williams
Dr. Anne D. Zimmerman
Dr. Gail Zimmerman
Mr. Michael D. Zimmerman
Mr. Mitchell Dean Zimmerman
Ms. Rhonda Sue Zimmerman

John Templeton Foundation—
Theology of Humility
March 1, 1990

I. Theology of Humility
  A. Centered in an Infinite God
    The Theology of Humility means we know so little and need to learn so much and to devote resources to research. The Theology of Humility is not man-centered but God centered. It proposes that the infinite God may not even be describable adequately in human words and concepts and may not be limited by human rationality. Perhaps God is not limited by our five senses and our perceptions of three dimensions in space and one dimension in time. Perhaps there was no absolute beginning and there will be no absolute end, but only everlasting change and variety in the unlimited purposes, freedom and creativity of God.

    Maybe God is all of time and space—and much more. The appearance of humankind on this planet may be said to have heralded the coming of a new quality encircling the earth, the sphere of the intellect. Then as we have used our intellects to investigate this mysterious universe, accumulating knowledge at an ever-increasing rate, there has come a growing awareness that material things are not what they seem; that maybe thoughts are more real and lasting than matter and energy.

    Perhaps this heralds a new quality, the sphere of the spirit. God may be creating not only the infinitely large but also the infinitely small; not only the outward but also the inward; not only the tangible but also the intangible. Thoughts, mind, soul, wisdom, love, originality, inspiration and enthusiasm may be little manifestations of a Creator who is omniscient, omnipotent, eternal and infinite. The things that we see, hear and touch may be only appearances. They may be only manifestations of underlying forces including spiritual forces which may persist throughout all the transience of physical existence. Perhaps the spiritual world, and the benevolent Creator whom it reflects, may be the only reality.

Presumably the sphere of the spirit may enclose not only this planet but the entire universe, and so God is all of Nature, is inseparable from it, and yet exceeds it. Perhaps it is mankind's own ego which leads us to think that we are at the center rather than merely one tiny temporal outward manifestation of a vast universe of being which subsists in an eternal and infinite reality which some call God. Maybe all of nature is only a transient wave on the ocean of all that God eternally is. Maybe time, space and energy provide no limit to the Being which is God. Likewise the fundamental parameters of the universe—the speed of light, the force of gravitation, the weak and strong nuclear forces and electromagnetism—would seem to pose no limits to the Being which is God.

B. Creative, Progressive

The Theology of Humility encourages change and progress. Science is revealing to us an exciting world in dynamic flux whose mechanisms are ever more baffling and staggering in their beauty and complexity. Scientists are learning to live and work with quantum uncertainty and complementarily, major discontinuities in evolution and baffling complexity in cellular differentiation. Yet scientists have turned these and other discoveries into opportunities, and many of them have expressed a new openness to philosophical and religious questions about life and the universe.

While science has generally responded favorably to change, the long history of religion is filled with the failures of thousand of religions. Perhaps these religions disappeared partly because their conceptions of God were too small or their practitioners were too inflexible to receive new revelations. A Theology of Humility proposes that maybe God is now providing new revelations in ways which go beyond any religion, to those who welcome the originality of the Creation and its continual surprises. For example, some theologians and scientists see tremendous possibilities for our future understanding of ourselves and our Creator through an integration of the new discoveries of science with many religious traditions—a new "theology of science."

Perhaps our human concepts of God are still tied to a previous century. The Twentieth-first Century after Christ may well represent a new Renaissance in human knowledge, a new embarkation into the concepts of the future. Persons now living can hardly imagine the small amount of knowledge and the limited concepts of the cosmos which man had when the scriptures of the five major religions were written. Perhaps old scriptures need new interpretations. The Theology of Humility seeks to build on the great theologies of the past and present and does not oppose any other theology. It welcomes the ideas and inspiring literature of all religions. But perhaps we should be open to the possibility of various new unprecedented religions where the revolutions in our conceptions of time, space and matter significantly shape our theology. Perhaps, while recognizing that God should not be thought of as impersonal, our names for God should be less heavily focused on personhood, since their usage favors man-centered concepts. The Creator seems to be both transcendent and immanent, accessible both by science and by prayer, ready to transform the lives of those who invite him in.

The Theology of Humility encourages thinking which is open minded and conclusions which are qualified with the tentative world "maybe." It encourages change and progress and does not resist any advance in the knowledge of God or of nature, but is always ready to rethink what is known and to revise the assumptions and preconceptions behind our knowledge. It is possible that through the gift of free will God allows us to participate in this ongoing creative process. Perhaps a prerequisite on our part is to look beyond our biases and our fears, our personal hopes and aspirations, to see the glorious planning and the infinite majesty of the Planner. Maybe we should also ask ourselves—whether we are students of the natural or spiritual worlds—to study and experience the intimate relationships between physical and spiritual realities in our own lives.

C. Diversity

The Theology of Humility does not encourage syncretism but rather an understanding of the benefits of diversity. Constant change would seem to be the character of our universe. Despite the cycles of day and night and the seasons, we may be learning that nothing really repeats. The pattern that unfolds with time may not close back upon itself. Maybe it moves ahead, upward a little at each turn. The evolution of our universe would seem to be vast in its conception, yet curiously experimental and tentative, a truly creative work in progress. Perhaps human beings, so late an appearance in this evolutionary process, have been given some creative role in seeking to understand and interpret awesome any mysterious processes which science only now begins to fathom. We suppose that our part might be likewise to conceptualize and experiment over a wise diversity of possibilities in the physical and spiritual worlds.

In the physical world perhaps we should reexamine the arrow of time. Thermodynamics has tended to provide a picture of irreversible movement from order to randomness as the universe "runs down." But this is the antithesis of what appears to happen in the evolution from simplicity to complexity which has occurred in the evolution of the universe and of life. It is as though apparently self-integrated units of the simplest matter exhibit powers which successfully oppose the trend to randomness and produce instead orderly events and structures. Where manmade structures decay, natural systems seem driven toward growth and toward greater diversity and complexity. Here perhaps we have one of the greater laws describing the nature of the cosmos. If only blind chance were involved in evolution, we might expect decay and disorder. Yet the end product of evolution up to this point, is a conscious being endowed with a remarkable brain and dominated by purpose. Perhaps the physical world can only be understood as the expression of the purposes of the Creator.

In the spiritual world perhaps diversity is also reflected in the variety of religions and in the multiplicity of denominations. It may be that this increasing diversity provides for a freedom and a loving and healthy competition without which there might be only lesser progress. Perhaps we should applaud the new research programs and the new organizations arising in each of the world's religions.

II. Encouraging Progress

The Theology of Humility does not rely on man-made institutions or governments, nor does it seek to influence them. Perhaps one of the greatest developments in human history has been the increasing possibility for the freedom of each individual to learn and grow and develop. The Theology of Humility seeks to improve the human condition by internal and spiritual sources rather than by external human governments or institutions. It encourages worship of the Creator rather than dependence upon government, and spiritual growth rather than human, social and political activities.

The Theology of Humility suggests that tremendous benefits could accrue from our greater understanding of spiritual subjects such as love, prayer, meditation, thanksgiving, giving, forgiving and surrender to the Divine will. It further suggests that since science is opening our eyes to the vast works of an Infinite Creator, science may also be applied to varieties of experimental and statistical study of these spiritual entities. It may be that we shall see the beginning of a new age of "experimental theology," wherein studies may reveal that there are spiritual laws, universal principles which operate in the spiritual domain just as some natural laws function in the physical realm. Perhaps we will discover that the sphere of the spirit is intensifying as God's evolving plans unfold and accelerate.

Perhaps research foundations and religious institutions should devote vast resources and manpower to these scientific studies in the spiritual realm, equal or greater in magnitude to those currently expended on studies in the physical realm. There could be enormous rewards in terms of increased human peace, harmony, happiness and productivity if we collected more evidence that the God of the universe had put us here on this planet to learn from and challenge each other and to act as channels to radiate God's love, wisdom and joy.

Recipients of the Templeton Prize for Progress in Religion

1973   Mother Teresa of Calcutta, founder of the Missionaries of Charity. She sees Christ in the "poorest of the poor" in what has become a world-wide ministry to the dying.

1974   Brother Roger, founder and Prior of the Taize Community in France. Taize communes have appeared all over the world, bridging between many denominations and languages.

1975   Sir Sarvepalli Radhakrishnan, who was President of India and Oxford Professor of Eastern Religions and Ethics. A strong proponent of religious idealism as the most hopeful political instrument for peace.

1976   H.E. Leon Joseph Cardinal Suenens, who was Archbishop of Malines-Brussels. A pioneer of the charismatic renewal and a strong proponent of Christian unity.

1977   Chiara Lubich, founder of the Focolare Movement, Italy which has become a worldwide network of over a million people in communes and private homes engaged in spiritual renewal and ecumenism.

1978   Professor Thomas F. Torrance, who was Moderator of the Church of Scotland. A leader in the new understanding of the convergence of Theology and Science.

1979   Rev. Nikkyo Niwano, founder of Rissho Kosei-Kai and the World Conference on Religion and Peace, Kyoto, Japan. A Buddhist world leader in efforts toward peace and understanding among religious groups.

1980   Professor Ralph Wendell Burhoe, founder and former Editor of *Zygon Journal,* Chicago, U.S.A. A leading advocate of an intellectually credible synthesis of the religious and scientific traditions.

1981   Dame Cecily Saunders, originator of the modern hospice movement, England. Pioneer in the care of the terminally ill by emphasizing spiritual growth and modern methods of pain management.

1982   Rev. Dr. Billy Graham, founder of the Billy Graham Evangelistic Association, U.S.A. He has preached the Christian gospel in over 50 countries, brought diverse denominations together and promoted respect for all peoples.

1983   Mr. Alexander Solzhenitsyn, U.S.A. Historical writer and novelist who has been an outspoken critic of totalitarianism and a strong proponent of spiritual awakening in the democracies as well.

1984   Rev. Michael Bourdeaux, Founder of Keston College, England, a research center for the study of religion in communist countries. He has been a fearless supporter of Christians in Russia.

1985   Sir Alister Hardy, who was founder of the Sir Alister Hardy Research Center at Oxford, England. An outstanding biologist, he also had a deep interest in man's spiritual nature. His work has demonstrated widespread religious experience in the British Isles.

1986   Rev. Dr. James McCord, who was Chancellor of the Center of Theological Inquiry, Princeton, U.S.A. A leader in spiritual education as President of Princeton Theological Seminary.

1987   Rev. Professor Stanley L. Jaki, O.S.B. Professor of Astrophysics at Seton Hall University, U.S.A. He has provided a reinterpretation of the history of science which provides a context for renewed belief in God in a scientific age.

1988   Dr. Inamullah Khan, Secretary-General, World Muslim Congress, Karachi, Pakistan. Proponent of peace within and between the world's religions.

1989   The Very Reverend Lord MacLeod of the Iona Community, Scotland. A leader for spiritual renewal in the Church of Scotland.
Jointly with
Professor Carl Friedrich von Weizsäcker of Starnberg, West Germany. Physicist and Philosopher, a strong voice for dialog between science and theology.

1990   Baba Amte, India. A learned Hindu scholar and philanthropist who has relieved the poverty of millions in rural India.
Jointly with
Professor Charles Birch, Sydney, Australia. Molecular biologist and strong proponent of process theology and environ-

1991   The Rt. Hon. Lord Jakobovits, London. Chief Rabbi of Britain, a leader in Jewish concern for medicine and especially medical ethics.

1992   Rev. Dr. Kyung-Chik Han, Korea. Pioneer in helping the Presbyterian church to become in only 30 years the largest Presbyterian denomination on earth.

1993   Charles Wendell Colson, founder of Prison Fellowship. A strong Christian force for change in the American prison system.

1994   Michael Novak, historical and theological scholar at the American Enterprise Institute for Public Policy. A powerful voice for re-emphasis of our rich religious and philosophical traditions.

1995   Dr. Paul Davies, currently Professor of Natural Philosophy at the University of Adelaide in Australia. He is a leading authority in expounding the idea of purpose in the universe and author of more than 20 books.

1996   Dr. William R. Bright, president and founder of Campus Crusade for Christ International.

1997   Pandurang Shastri Athavale, founder and leader of a spiritual self-knowledge movement in India that has liberated millions from the shackles of poverty and moral dissipation.

TEMPLETON FOUNDATION PRESS

Worldwide Laws of Life
200 Eternal Spiritual Principles
John Marks Templeton
Wisdom drawn from major sacred Scriptures of the world and various schools of philosophical thought, as well as from scientists, artists, historians, and others. Thoughtfully arranged in a format that can be used as a resource for discussion groups.
ISBN 0-8264-1018-9 $24.95, hb
ISBN 1-890151-15-7 $14.95, pb

How Large is God?
The Voices of Scientists and Theologians
Edited by John Marks Templeton
A new collection of essays suggesting that our definition of God is too narrow. Includes chapters by Herbert Benson, Martin Marty, Owen Gingerich, Freeman Dyson, F. Russell Stannard, Howard Van Till, Robert Russell, John Barrow, and Robert Herrmann.
ISBN 1-890151-01-7 $22.95, hb

Is Progress Speeding Up?
Our Multiplying Multitudes of Blessings
John Marks Templeton
Loaded with statistics, charts, and photographs that illustrate that the state of the world is actually getting better. A timely resource book for optimists.
ISBN 1-890151-02-5 $19.95, hb

Golden Nuggets
from Sir John Templeton
This collection of inspiring thoughts from Sir John touches on the things that make life meaningful: happiness, love, thanksgiving, forgiveness, positive thinking, and humility.
ISBN 1-890151-04-1 $12.95, hb

Is God the Only Reality?
Science Points to a Deeper Meaning of the Universe
John Marks Templeton and Robert L. Herrmann
Examining discoveries in fields such as particle physics and molecular biology, the authors address the great paradox of science that the more we learn, the more mysterious the universe becomes.
ISBN 0-8264-0650-5 $22.95, hb

Evidence of Purpose
Scientists Discover the Creator
Edited by John Marks Templeton
A thought-provoking collection of essays by respected scientists who explore new developments in their fields and the consequent theological implications.
ISBN 0-8264-0649-1 $24.50, hb

For a complete listing
of titles available
visit our web site
www.templeton.org/press

To order books
call (800) 621-2736

Templeton Foundation Press
Five Radnor Corporate Center
Suite 120
100 Matsonford Road
Radnor, Pennsylvania 19087

LET THEM SEE YOUR TALENT!

LITTLE RHINO #4 NOW AVAILABLE!

Dear Reader,

I was just about Little Rhino's age the first time I spent a night away from home for a baseball tournament. I was filled with excitement and anticipation as I packed my bags for my overnight adventure. I was nervous because I wasn't going to have my family around. Would I get homesick? But at the same time, I was thrilled to have the opportunity to spend a night away from home in a hotel with all of my teammates. I was excited to play my favorite sport in a stadium I had never visited. I remember packing a ton of snacks, a deck of cards, and my baseball gear.

As I got older and made it to the minor and major leagues, I was never nervous to travel. I was curious to see what each city had to offer. History was one of my favorite subjects in school, so exploring the cities on an off day or before I had to report to the ballpark made traveling even more fun. Every field was a little different. Every team had its own style. I enjoyed staying in hotels with my teammates because they felt like family to me. We would hang out, eat together, and really get to know each other. It still is one of my favorite parts of the game!

Krystle Howard

LITTLE

Rhino

by **RYAN HOWARD**
and **KRYSTLE HOWARD**
WITHDRAWN

● **BOOK FIVE** ●

THE AWAY GAME

SCHOLASTIC INC.

To Alexandria. The world is yours.

—R.H. & K.H

Text copyright © 2017 by Ryan Howard
Illustrations © 2017 by Scholastic Inc.

This book is being published simultaneously in hardcover by Scholastic Press.

All rights reserved. Published by Scholastic Inc., *Publishers since 1920.* SCHOLASTIC, SCHOLASTIC PRESS and associated logos are trademarks and/or registered trademarks of Scholastic Inc.

The publisher does not have any control over and does not assume any responsibility for author or third-party websites or their content.

No part of this publication may be reproduced, stored in a retrieval system, or transmitted in any form or by any means, electronic, mechanical, photocopying, recording, or otherwise, without written permission of the publisher. For information regarding permission, write to Scholastic Inc., Attention: Permissions Department, 557 Broadway, New York, NY 10012.

ISBN 978-1-338-05234-3

10 9 8 7 6 5 4 3 2 1   17 18 19 20 21

Printed in the U.S.A.   40
First printing 2017

Book design by Christopher Stengel

· CHAPTER 1 ·
An Honor

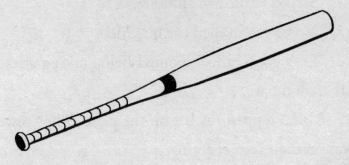

Rhino! Rhino! Rhino!"

Little Rhino glanced at his teammates, who were standing inside the dugout fence, chanting his name. He nodded, but tried hard not to smile.

Rhino gripped his bat and strode to the batter's box. He took a couple of easy swings and glared at the pitcher. The pressure was on!

Bases loaded. Two outs. The Mustangs trailed by two runs in the bottom of the sixth inning. This was their last at bat.

The Bears' pitcher squinted as he read the catcher's signals. Then he leaned forward and fired

a sizzling fastball. It looked wide, and Rhino let it go by.

"Strike!" called the umpire.

"No batter!" yelled the infielders.

"Let's go, Rhino," shouted Bella, taking a lead off second base.

Rhino tapped his bat on the plate. The crowd was on their feet, cheering.

"Ball," said the umpire as the catcher leaped high to grab the next pitch.

First place in the league was on the line in this game, and the showdown between Rhino and the pitcher was shaping up to be a classic. Rhino led the league with five home runs, and the Bears pitcher had the best record.

Rhino had belted a two-run homer in the first inning, but he'd struck out twice since then. The Bears had played steady baseball all season. Rhino's Mustangs had come into the game on a hot streak, with five wins in a row. Whichever team won today would be in first place.

Rhino took a quick look at the other base runners: Carlos on third base and Dylan on first. Both were bouncing on their toes, ready to sprint.

Their teammates in the dugout chanted again. "Rhino! Rhino!"

Here came the pitch. Rhino took a massive swing.

The Bears yelled as the catcher safely caught the ball.

"Strike two!"

Rhino's thinker told him to relax. *You've been in pressure situations before. Just meet the ball.*

Rhino looked out at the scoreboard in center field. BEARS 7, MUSTANGS 5.

He took a deep breath.

Whack! Rhino felt his muscles surge as he clobbered the next fastball. The ball streaked toward the scoreboard, but Rhino didn't watch it. He raced to first base as the crowd whooped.

Rounding first, Rhino saw the Bears' center fielder watching helplessly as the baseball flew over

his head for a home run. Rhino leaped with both hands up as he stamped on second base, then continued running.

The Mustangs won. They were in first place!

Carlos, Bella, and Dylan waited for Rhino as he ran toward home. They pounded his back and yelled as the rest of the Mustangs ran from the dugout.

"Two blasts!" shouted Bella. That was a first. Rhino had never hit two homers in one game before.

The Bears' pitcher looked glum as the teams shook hands.

"Great game," Rhino told him.

"Nice hit," the pitcher mumbled. "But we'll get you next time."

"Good luck until then," Rhino replied. "Keep that arm loose."

Coach Ray gathered the Mustangs in the dugout. "I'm very proud of this team," he said. "Winning six straight games is tough. But what I'm most proud of is your hard work and sportsmanship."

Rhino clapped and the others joined in. "Thanks, Coach," he said. "You're a great leader."

Bella nudged Rhino with her elbow and smiled.

"Winning today gives us a real honor," Coach said. "Since we're in first place, we'll be representing our league in a special tournament next weekend. It's an exhibition, so it won't affect how we're doing in this league. It is basically a practice game. But it will be an exciting trip and gives us a chance to test ourselves against some very strong teams."

Wow, Rhino thought. *That sounds like a Major League honor.* He couldn't wait.

Coach told the players that he'd understand if not everyone decided to make the trip. They'd be traveling by bus to the state capital and staying overnight in a hotel. If they won their first game, they'd play in the championship game the next day.

Rhino gulped. Two nights in a hotel? Away from C.J. and Grandpa James? Rhino had never slept anywhere except Grandpa James's house. Not

even for a sleepover at his best friend Cooper's. And Rhino's brother, C.J., was always there with him.

"This will be so great," Bella said to Rhino as they left the dugout.

Rhino nodded. That was easy for Bella to say. Her father would be along for the trip. Coach Ray was her dad.

Rhino felt a tap on his shoulder. He turned to see Dylan's wise-guy smile.

"Another great win, huh?" Rhino said.

Dylan shrugged. "Of course. But we'll see how well you do in the big-time tournament next weekend," he said. "Things are different at the capital. *Intense* competition."

Dylan was always trying to stir up trouble. *Just ignore him,* Rhino's thinker said. "How would you know?" Rhino asked.

"I've been in plenty of big sports events," Dylan said.

"Like what?" Rhino replied. Dylan didn't have

any more experience than Rhino did. They were both playing on a real team for the first time this season.

"Too many to name," Dylan said.

"Name one."

Dylan changed the subject. "Be sure to bring your teddy bear," he said. "Being away from home overnight is going to be scary. For you."

"Why should I be scared?" Rhino asked. He *was* feeling uneasy about it. But he wasn't going to let Dylan know that.

"Believe me, you'll be afraid," Dylan said. "No big brother around. No grandfather."

Rhino started to walk away. "Get lost, Dylan. You don't know what you're talking about."

"Yeah, lay off," Cooper said, stepping over. He jutted his head toward the gate. "Let's go, Rhino."

"Great homer," Cooper said when they were out of Dylan's earshot. "And great news: My dad is

going on the trip with us. You and I can room with him at the hotel."

That made Rhino feel a little better. Cooper's father was always kind and supportive to him.

Then Rhino had an idea. Maybe Grandpa James could be there, too! Having Grandpa along would make it the best trip ever.

Rhino felt that same surge of excitement he'd felt when he hit the game-winning homer. "State capital, here we come!"

· CHAPTER 2 ·
Facing the Fear

Rhino ran all the way home. He couldn't wait to tell Grandpa James about the trip. And about his two home runs!

"Grandpa!" he called as he hurried through the back door. "Wait until you hear!"

Grandpa James caught Rhino in a bear hug and laughed. Rhino told him the news. "Cooper's dad is coming along," he said. "Can I go? Can you be there, too?"

Grandpa poured Rhino a glass of milk. "First of all, congratulations on the game," he said. "And of course you can go on the trip. You've

worked very hard in baseball, and in your schoolwork. So you've earned it."

Rhino smiled. "What about you?"

Grandpa patted Rhino's shoulder. "Next Saturday evening is C.J.'s science fair," he said. "I'd love to go to your tournament, but I promised your brother weeks ago that I'd be at the fair."

"Can you drive to the tournament after?" Rhino asked.

Grandpa smiled. "I don't think so. The capital is nearly three hours away. We would get there very late at night."

Rhino rubbed his chin. "I guess I'll be okay," he said softly.

"Sure you will," Grandpa said. "But it's up to you. If you'd rather stay home, I'll understand. And so will your coach. He phoned me last night."

"He did? So you knew about the tournament before I did?"

"Of course," Grandpa said. "Coach Ray didn't want to tell the players before today's game. But he

needed to make sure the parents knew. I told him I believed you'd want to play, but that it would be your decision."

Rhino stared at his glass of milk. He did want to play. And he certainly didn't want his teammates to think he was afraid to spend a night away from home. Dylan would never let him hear the end of that!

"Think it over," Grandpa James said. "But you'll need to make a committed decision by Monday. Once you decide, there will be no changing your mind."

Rhino nodded. His thinker told him to be brave. *This will be fun! And what a great chance to show your skills against the best players in the state.*

"I'm going," Rhino said firmly. "I'll be okay."

"I know you will," Grandpa said. "You've done a lot of things to be proud of this spring. This will be another one, no matter what happens in the tournament."

Rhino felt better. He finished his milk and made himself a peanut-butter-and-jelly sandwich. His favorite meal!

When C.J. came home, he and Rhino went out to the yard to have a catch. C.J. was in seventh grade—four years ahead of Rhino. He'd been playing on sports teams for several years, and was the starting shortstop for his middle-school team.

"I heard about the tournament," C.J. said. "I wish I'd had an opportunity like that when I was your age!"

"It'll be very cool," Rhino said. But he didn't sound as enthusiastic as before.

"I'm sure it will," C.J. said. "Something bothering you?"

C.J. fired a high throw and Rhino had to leap for it. Rhino nabbed the ball, spun around, and tossed it back.

"Nice grab," C.J. said.

Rhino caught another throw. He tossed it high in the air. C.J. circled under it and made the easy catch.

"So?" C.J. asked.

Rhino shrugged. "The games don't worry me," he said. "But . . ."

"Being away from home does?"

"A little."

C.J. nodded. "I was ten the first time I did an overnighter," he said. "A year older than you are. Remember? I went camping with Robert's family."

"I think so." Rhino threw a hard grounder and C.J. scooped it up.

"I was so excited about it," C.J. said. "Fishing, hiking, cooking hot dogs over the fire. It was going to be so cool."

"Wasn't it?" Rhino caught the ball and held it. He took a few steps toward C.J.

"It was," C.J. said. "All those things were great. We set up a couple of tents right on the shore of the lake. The moon was full that night and you could hear a million crickets chirping, and every once in a while a fish would jump. It *was* cool."

Rhino frowned. "So you weren't scared at all?"

C.J. laughed. "I didn't think I was. Then Robert and I crawled into our tent. He fell asleep in about two seconds. I just laid there, thinking about how far away I was from Grandpa James. I'd never slept in any house but this one. What if something happened? What if a bear came by, or I got lost in the woods somehow?"

"So you *were* scared."

"I was petrified." C.J. nodded toward the back steps, and he and Rhino sat down. "After about a half hour I woke Robert up. Not to tell him I was scared. I made up some excuse, like I needed a snack."

Rhino laughed. "What did he do?"

"He rolled over and went back to sleep. So I grabbed a snack. I had a candy bar in my knapsack. That helped for a minute."

Rhino's thinker told him to remember that. *Bring snacks!*

"Every sound made me more scared," C.J. said.

"I opened the tent flap and looked out. It was dark, but I could see shapes in the moonlight."

"Wow."

"Yeah." C.J. pounded his fist into his mitt. "I just watched the lake for a long time. And I remember thinking, 'Robert's parents are right there in the other tent.' I started to feel safer. And it was a beautiful night. I was never *not* scared that night, but it got easier. Sooner or later I fell asleep. Before I knew it, morning arrived."

"It always does, huh?"

C.J. put his arm around Rhino's shoulder. "Every time," he said. "So, enjoy the tournament. Don't complicate things by worrying about being away from home. You'll get through that. Just like I did."

· CHAPTER 3 ·
Rhino's New Skill

The Mustangs were excited but nervous at practice that week. None of them had ever played in a game outside of this league.

"There's enough pressure here, where we know everybody on the other teams," Bella said to Rhino as they tossed a ball around before practice began. "What will it be like on Saturday against a team from another part of the state?"

"Fun, I hope," Rhino said. "We're a good team. We'll do fine."

The game didn't worry Rhino at all.

"This tournament will be a big step for us," Coach Ray said. "We need to be very sharp, so let's work on our fielding skills."

Rhino jogged to first base. It was one of the toughest positions in baseball, and he was proud to play there. This baseball field was beginning to feel like his second home.

Coach Ray stood near Rhino, just behind the base. "Let's work on catching the ball," he said.

Rhino nodded. *But I'm* good *at catching the ball,* he thought. He'd only dropped one all season.

Coach smiled. "You look confused, Rhino. What I mean is, let's think about how to stand, depending on who's throwing it to you."

Rhino wasn't sure what Coach meant.

"You've been off-balance a few times in the games," Coach continued. "Remember in the fourth inning the other day? You had to lunge way over on that throw from Cooper."

"I remember," Rhino said. "But I did catch it."

Coach waved Rhino aside and stepped onto the infield. "Watch me," he said.

Coach threw the ball to Sara at third base. Then he ran to the bag and faced her, waiting for the ball to return. Sara's throw was a little wide, but Coach reached it easily, keeping his foot on the base.

"Keep watching," Coach said. He bounced the ball to Cooper at shortstop, then to Carlos at second base. Coach made the easy catches both times.

"Notice anything?" Coach asked.

Rhino shrugged. "You did everything right. You didn't stretch way out unless you had to. Kept your foot on the base."

"Right," Coach replied. "Just like we've been working on all season. Anything else?"

Rhino shook his head. "I don't think so."

"Watch again," Coach said. He threw the ball to Sara at third.

"Now watch my shoulders," Coach said. He made the catch and turned to Rhino. He held up

his hand and pointed at Sara like an arrow. "See how my shoulders were squared up with her?

Coach went through the same routine with throws from Cooper and Carlos. He turned to face the thrower directly each time.

"Whoever is throwing the ball to me, I want my shoulders squared up to him or her. The middle of my chest is facing straight at that player."

"I get it," Rhino said. *Another new skill,* his thinker told him. Learning to play first base was hard, but little by little he was becoming a standout.

"Try it," Coach said.

Rhino took throws from each of the infielders. *Square up,* he told himself each time. It was easy to remember.

"Batter up!" Coach called.

Bella batted first.

"Just make contact with the ball," Coach said. "We don't need home runs today, just solid hits and nice, steady throws."

Rhino bounced up and down a few times, then got in his fielding stance.

Bella smacked the ball hard, just to the side of second base. Carlos fielded it and pivoted, then fired the ball toward Rhino.

Rhino faced Carlos. From the corner of his eye he could see Bella sprinting toward the base.

Carlos's throw was high and way off line. Rhino reached across but couldn't get to it without leaving the base. He made the catch, but Bella was safe.

"Good hustle, Bella," Coach said. "And good work, Rhino. You did what I said, and you kept the ball in play."

Bella grinned at Rhino. "Too quick for you," she said.

The next batter lined the ball toward shortstop, and Cooper grabbed it on one bounce. *Square up,* Rhino thought. He made the catch easily.

Things went pretty well all afternoon. Rhino remembered what to do on all the routine plays.

It wasn't so easy on tricky ones like double-play attempts or when Dylan bunted and the throw came from the catcher.

"Great work," Coach said to him after practice. "Like everything else about playing first base, it will get easier after a while. Just relax. Soon you'll square up automatically. You won't have to think about it."

"Thanks, Coach." Rhino gathered his bat and glove and sweatshirt from the bench. Then he heard a familiar voice.

"It's going to be a *long* weekend," Dylan said, poking his head into the dugout. "No one to tuck you in at night or sing you a lullaby."

"Very funny," Rhino replied. "Why don't you grow up, Dylan? You always want to bully somebody."

"Yeah," said Bella. "Grow up, Dylan. Rhino isn't afraid of anything."

Dylan laughed. "Then he shouldn't be bothered by a little teasing."

"I'm not," Rhino said. He turned to Bella. "See you tomorrow in school."

As he walked home, Rhino tried to stay positive. He'd learned a new first-base skill today. The tournament would be exciting and fun.

But one thought kept coming back to him. *Two nights away from home.* Even his thinker didn't have a good response to that.

Just stay focused, his thinker told him. *Just have fun.*

Rhino knew that wouldn't be easy. He stopped and looked back at the baseball field. He was comfortable here. He hit home runs and made great plays at first base. And his teammates considered him a leader.

Everything would feel different on Saturday. He'd be away from home for the first time. And he'd also be away from his *second* home.

· CHAPTER 4 ·
Baseball Jokes

Rhino rushed home from school on Friday. The bus was scheduled to leave for the state capital at 4:00 P.M., and he wanted to spend time with Grandpa first.

Rhino carefully folded his Mustangs uniform and some extra clothes. "I'll wear my cap," he said to Grandpa, proudly placing the blue cap with the big white *M* on his head.

"You look like an all-star," Grandpa said with a grin.

Rhino went to the kitchen and grabbed a loaf of bread. He counted out twelve slices, then took

the peanut-butter and jelly jars from the cabinet. He was spreading peanut butter on the second slice when Grandpa came in.

"That's a lot of sandwiches," Grandpa said.

Rhino counted off the number on his fingers. "Dinner tonight; breakfast, lunch, and dinner tomorrow; breakfast and lunch on Sunday."

Grandpa laughed. "You don't need to bring your own food," he said. "That will be provided."

"Oh." Rhino looked at what he'd already prepared. "Maybe I should bring a few sandwiches anyway."

"Good idea," Grandpa said. "How about two?" He started to put the unused slices back in the bread bag.

"Midnight snacks," Rhino said. "I get hungry!"

"I know," Grandpa said. "You and C.J. have mighty big appetites."

Grandpa drove Rhino to the baseball field, where a bus was waiting. Most of the team members were already on board.

Cooper's dad was waiting outside. He shook hands with Rhino and Grandpa James. "Cooper's saving you a seat in there," he said.

Rhino gave Grandpa a hug. His eyes were stinging, but he smiled. He waved hard to Grandpa as the bus pulled away.

Several parents were on the bus. They sat up front. Rhino felt better as soon as he took his seat next to Cooper. Bella switched on some music and they sang along as the bus headed down the highway.

Everyone on the team had decided to make the trip. There was a lot of joking around and talk about how exciting the games would be. Rhino forgot all about being worried.

"Why isn't Cinderella a good shortstop?" Bella asked.

No one knew the answer to Bella's joke.

"Because she runs away from the ball!"

"I've got one," Carlos said. "Why can't a zebra ever reach first base?"

"Because he can't hold the bat?" Rhino asked.

"No," Carlos replied. "Three stripes and you're out!"

Most of the jokes were silly after that, but everyone was laughing and acting goofy. *Almost* everyone.

Rhino glanced back. Dylan was sitting alone several rows behind him, staring out the window. He wasn't laughing at all.

Rhino nudged Cooper. "What's with Dylan?"

Cooper took a quick look. "Maybe he's bus sick."

"It's been a smooth ride," Rhino said. He kneeled on the seat and said, "Hey, Dylan."

Dylan looked up.

"You all right?" Rhino asked.

Dylan nodded. "Just thinking about the tournament," he said.

"Think about it later," Cooper said. "Have some fun."

Dylan pulled his Mustangs cap low over his eyes and shook his head.

"Let him be," Rhino said. At least Dylan wasn't making fun of anyone, as he often did. But Dylan didn't look serious. He looked worried.

After a while, Rhino went back and sat across the aisle from Dylan. Dylan glared at him.

"How's your arm?" Rhino asked. Dylan was scheduled to pitch for the Mustangs tomorrow.

Dylan bent his elbow and rubbed his bicep. "It feels great," he said. He turned toward the window again. Then he asked, "You?"

"I'm ready," Rhino said. "A little nervous . . . about the tournament. I guess that's natural."

Dylan hunched up and said, "Yeah."

Rhino could tell that Dylan didn't feel like talking.

After dinner at a pizza restaurant, the team checked into their hotel. "Nice room," Cooper said. There were two beds and a cot. Rhino set his knapsack on the cot and looked around.

"Do you want a bed?" Cooper asked.

"You and your dad can have those," Rhino said. The cot looked cozy. The beds were bigger than his at home.

"Look how high up we are," Cooper said, pointing out the window. Their room was on the eighth floor.

Rhino looked out, too. The city was much bigger than their hometown—lots of tall buildings and busy streets. "I wonder where the baseball field is," he said.

Cooper pointed to a stadium about three blocks away. Bleachers wrapped around the entire baseball field, and colorful pennants blew in the breeze. It almost looked like a professional stadium.

They heard laughing out in the hall, then a knock on the door. Carlos, Bella, and two other teammates came in.

"Can we play some games in here?" Bella asked, holding up a box.

"Sure," said Cooper's dad. He turned on the

television to watch a baseball game. The kids crowded around the small table.

"I love overnight trips," Carlos said.

That surprised Rhino. First of all, when had quiet Carlos ever been on a trip like this? And second, Carlos was usually nervous about things like baseball games and tests. Why was he so relaxed tonight?

"My church choir goes on trips a couple of times a year," Carlos said. "We sing at other churches. But the most fun is at night—telling jokes in the hotel rooms, making a mess with chips and sodas."

"Do your parents go on the trips with you?" Rhino asked. He immediately wished he hadn't said that. Why else would he ask unless he was afraid of being away from Grandpa James?

"Sometimes they do," Carlos said. "Sometimes not."

Rhino nodded. He noticed that Bella was giving him a small smile. She winked.

"Where's everybody else?" Cooper asked.

"In another room, doing the same thing we are," Carlos said. "Except Dylan. He went right to bed."

"Is he all right?" Cooper's dad asked.

"I think so," Carlos said. "He just said he was tired. Hey, why does it take so long to run from second base to third?"

"It doesn't," Rhino replied.

"It's a joke," Carlos said. "It takes so long because there's a short stop between the bases."

Rhino groaned. He tossed a pretzel at Carlos. Then he took a handful of chips and laughed. For the next hour he had a blast with his friends.

"Lights out in ten minutes," Cooper's father announced. The others went back to their rooms.

Rhino lay awake for a long time. He wasn't used to the sounds of the city. He heard a siren in the distance, and a truck beeped down below.

Tomorrow would be a big day. That stadium looked awesome.

Rhino missed Grandpa James. He wished he could tell him that. He missed C.J., too.

Rhino remembered C.J.'s story about camping out. The sights and sounds outside this hotel room were much different from the tent by the lake. But Rhino felt just as lonely as C.J. had.

Then he remembered something else about C.J.'s trip. *Have a snack,* his thinker told him.

Rhino quietly opened his knapsack. He unwrapped a PB&J sandwich. It was his favorite meal.

It made him feel just a little bit better.

CHAPTER 5

Dealing with the Nerves

Rhino yawned in the elevator as the team went down to the hotel restaurant for breakfast. He'd slept, but not as much as usual. Twice he'd woken up not knowing where he was. It made him jump.

For once, Rhino didn't feel hungry. He ordered two eggs and a slice of ham, but he stared at the food for a few minutes before taking a bite.

"Not cooked the way you like it?" Cooper asked.

"It's fine," Rhino said. He took another nibble. Grandpa made much better scrambled eggs. Better toast, too.

Cooper's dad's phone rang. "Hello?" he said.

"Oh, hi! Sure, he's right here." He reached across to Rhino. "It's for you."

"Me?" Rhino took the phone. "Hello?"

"Good morning, Little Rhino!" It sounded so good to hear Grandpa's voice on the other end. Rhino forgot all about being homesick. He told Grandpa about the fancy hotel room and the view of the city. "We laughed a lot last night," he said. "We played games and joked around. It was cool."

"That's wonderful," Grandpa said. "There's nothing like bonding with your teammates. Now, you have a great game today. Can't wait to hear about it."

"Wish you could see it!"

Rhino dug into his eggs. They tasted much better. He spread his toast with butter and jelly and drank his entire glass of orange juice.

Coach Ray stood up and tapped a fork on his plate to get everyone's attention. "Let's get suited up and meet in the lobby in fifteen minutes," he said. "We'll have batting practice at the field this

morning. Game time is eleven o'clock. I want every-body looking sharp. Shirts tucked in. Caps on straight. We want to look like we belong here!"

Rhino pulled the brim of his cap to make sure it was straight. Then he rushed upstairs to put on his uniform.

A team in red-and-white pinstriped uniforms was leaving the field as the Mustangs arrived. Their jerseys said CARDINALS.

"That's the team from the capital league," Coach said. "Our opponents this morning."

Some of the Cardinals looked much bigger than any of the Mustangs, Rhino thought. But maybe it was his imagination. They were all the same age—third- and fourth-graders.

The grass was thick, and it smelled fresh—as if it had just been mowed. The base paths were marked with straight white lines. The score-board had advertisements for restaurants and other businesses.

"Big time," Cooper said.

Rhino nodded. The outfield fence looked like it was a long way off. "Big *field*," he said.

"It's exactly the same size as ours," Coach said. "It just *seems* bigger."

It felt strange to be on a field surrounded by tall buildings. The field back home was in a large town park. The only tall things nearby were trees.

Rhino kneeled in the on-deck circle as others took their swings. Dylan waited nearby. He was much quieter than usual.

"Nice stadium, huh?" Rhino said.

Dylan shrugged.

"You sleep okay?" Rhino asked.

"Like a ba—" Dylan stopped short. "Like a puppy," he said.

Rhino winced. Dylan had started to say "Like a baby."

"Right," Rhino said. "It took a while to fall asleep, but . . ."

"Me too," Dylan replied. "Actually, puppies don't sleep all that soundly. Not ours, anyway."

"You have a puppy?" Rhino asked.

Dylan wiped his mouth with the back of his wrist. "A Labrador. Ten weeks old. He misses his mother, so he cries at night when we're trying to sleep."

"That's too bad," Rhino said. "It would break my heart."

Dylan shrugged. "I usually pick him up and put him on the foot of my bed," he said. "That stops him from crying, but it's a bad habit to get into. My dad says to just let him cry, otherwise he'll never get over it."

On the field, Cooper bobbled a ground ball and let it slip away. A minute later, Bella dropped an easy fly ball in right field.

"I think everybody's nervous," Rhino said. "We'd better settle down in time for the game."

Dylan let out his breath in a huff. "You're up," he said. "See if you can reach that fence."

Rhino picked up his bat and stepped toward

the plate. Then he turned back. "What's his name?" he asked Dylan.

"Who?"

"The puppy."

"Bruiser," Dylan said. "But he isn't tough at all."

Coach Ray was pitching. His throws were straight and fast, and Rhino lined the first two right up the middle.

"Let's see some power!" Carlos called from second base.

Rhino fouled off a couple of pitches. *Relax your shoulders*, his thinker said. *Smooth, steady swings.*

The next pitch was over the heart of the plate. Rhino whacked it and watched it soar high over right field. Bella didn't even move from her spot. She just turned and waved good-bye to the ball.

"Here we go!" Cooper called. "Just like back home!"

Rhino hit another long one, which cleared the fence near the scoreboard. Soon everyone seemed looser—calling encouragement to one another and

catching every ball that came their way. Now it seemed like a practice at their own field.

"Look around," Coach said as they gathered near home plate. "The distance between the bases is the same as at our field. The pitcher's mound is the same distance away. And the outfield fences are the same height, too. Just think of this as another game and we'll do fine."

The Mustangs put their hands together in a circle and chanted, "One . . . two . . . three . . . let's go!"

· CHAPTER 6 ·
Extra Hot

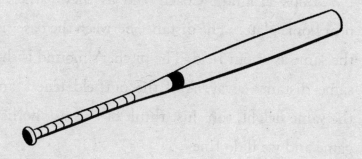

The Cardinals were as sharp as their uniforms. Their left-handed pitcher had a blazing fastball, and he struck out Cooper and Bella in the top of the first inning.

Rhino gripped his bat as he took his place in the on-deck circle. He was the Mustangs' cleanup hitter. But if Dylan struck out, too, there'd be no one on base to "clean up."

Dylan hit a weak grounder to the first baseman, who scooped up the ball and stepped on the base before Dylan could get there.

"Defense!" Rhino called. He set his bat in the rack, grabbed his glove, and ran onto the field.

Dylan walked slowly to the mound, looking very serious. He threw several warm-up pitches as Rhino tossed a ball "around the horn," firing it to the other infielders.

The bleachers were nearly full, and music played from the announcer's booth while the Mustangs warmed up. Bright sunshine and a light breeze made for perfect baseball weather. Rhino felt great.

Dylan had trouble settling down. He walked the first batter on four pitches.

"No problem, Dylan!" Rhino called.

"Get this next one!" yelled Bella. All the other Mustangs yelled encouragement, too.

Rhino stayed close to the base so the runner wouldn't take a big lead. The next batter swatted the ball deep into the gap between third base and shortstop. Cooper chased it down and made a

backhand grab. It was too late to get the runner at second, so Cooper threw the ball across the diamond to Rhino.

The throw was high and to Rhino's left. He leaped for it and made the catch. In the same motion, he swung his glove toward the batter, tagging him just before he reached the base.

"Out!" called the umpire.

"He's going for third!" Dylan called.

Rhino pivoted. The runner was more than halfway to third base. Rhino threw. The runner slid. Sara tagged him.

"Safe!" was the call.

"Square up, Rhino," called Coach Ray.

Rhino nodded. Would that have made a difference? Probably not. But he needed to remember that tip. He'd been off-balance when the throw came from Cooper. The baserunner had made a risky move in going for third, but it paid off.

The Cardinals' pitcher stepped up to the plate.

He was tall and limber, with curly black hair sticking out the back of his helmet.

"Let's go!" came a cry from the Cardinals' dugout. "Bring him home."

Dylan threw a couple of quick balls. Both were high and outside.

"No batter!" Rhino called. "Just throw strikes." He held up his index finger and called to his teammates. "One out!"

Every Mustang was surprised when the batter bunted the ball up the first baseline. Rhino raced to the bag and planted his foot. The catcher grabbed the ball with his bare hand and threw.

Rhino could barely see the ball as it streaked toward him behind the batter. He lifted his glove and somehow made the catch for the second out. But the other runner scored easily.

"Nice bunt," Rhino mumbled as the batter trotted past him to the dugout.

"Thanks," the kid said without looking back.

The next hitter lined the ball straight at

Rhino. He made the catch and ran to the dugout for his bat.

He'd never heard his name announced before, so it was a big thrill as he stepped to the plate.

"Leading off for the Mustangs: Number six, Ryan Howard."

The Mustangs' parents cheered loudly in the bleachers. Everyone else clapped politely.

Rhino toed the dirt with his right foot and let out his breath. Even though he was a lefty, he felt comfortable batting against another left-hander. *Bring it on,* he thought.

The first pitch blazed past for a strike.

Whoa, Rhino thought. *This guy definitely has some heat.*

He took a wild swing at the next one and didn't come close. *Settle down,* his thinker said. *Find the ball and meet it.*

The next pitch was just outside, but waist high and hittable. *Yes,* Rhino thought as his swing met the ball. He raced toward first base.

The spectators stood and gasped. The ball was soaring way out of the park. But it was outside the right-field foul pole. Rhino shook his head and trotted back.

"Straighten it out!" Cooper called. "One more like that."

Rhino's heart was beating fast. He eyed the pitcher. Got set. Swung!

"Strike three!" called the umpire.

Rhino walked swiftly back to the dugout.

The next two Mustangs struck out, too. And all three Mustangs made quick outs in the third.

"He's got a no-hitter going," Bella said as the Mustangs came to the bench in the top of the fourth.

"Worse than that," Rhino said. "It's a perfect game so far. The pitcher hasn't given up any hits or walks and nobody has made a fielding error."

But Dylan was pitching well, too. The Mustangs still trailed by only a run.

When Cooper led off with a walk, Rhino felt another surge of excitement. Unless there was a

double play, he'd bat this inning. He could tie the game or give the Mustangs the lead.

Bella grounded out. Dylan struck out.

"Don't leave him stranded," came a call from the Mustangs' dugout.

Rhino nodded. He'd been studying the Cardinals' pitcher. His pitches were just as fast as before, but they weren't moving much. Most of them were straight down the middle. Home-run pitches.

I'm blasting this one out of the park, Rhino thought. *Say bye-bye to that no-hitter and the lead!*

· CHAPTER 7 ·
A One-Run Margin

The Cardinals pitcher was wary. He'd seen Rhino's power in that long foul ball the last time, so he kept his pitches low and inside.

The count ran full. Three balls and two strikes. Cooper took a big lead off first base.

Bam! Rhino smashed the ball and it lined toward right field. The second baseman leaped and touched the ball with his glove, but it still rolled into the outfield. Rhino was safe at first, but he hadn't driven in a run. Cooper was stalled at third.

An easy pop fly ended the inning. CARDINALS 1, MUSTANGS 0.

"We've got two more at bats," Cooper said as they took the field. "Hold them scoreless. We can still win this thing."

At least Rhino had broken up the no-hitter. That might give the other Mustangs some confidence at the plate.

But the Mustangs went three up, three down in the fifth. Still, Dylan kept the Cards in check. It remained a one-run game as the Mustangs came up to bat in the sixth.

The first batter struck out, but that brought up the top of the Mustangs' order.

"Let's go, Coo-per!" chanted the players on the bench. Five quick claps, then "Let's go Coo-per!" again.

Rhino sipped from a water bottle. The Cardinals were tough, but Rhino was sure his team could win. *Just get on base, Cooper,* he thought. *I'll bring you home this time.*

Cooper came through. He smacked a line drive over the shortstop's head and it fell safely into left field. Bella followed with a single, too.

"Do it for the puppy," Rhino said to Dylan.

Dylan smirked. He popped a perfect bunt in front of the plate, moving Cooper and Bella to second and third. Dylan was out, but he'd done his job. Rhino held out his hand and Dylan slapped it as he trotted to the dugout.

"Rhino! Rhino! Rhino!"

A solid single would tie the game or give the Mustangs the lead. A home run would leave no doubt.

Rhino tapped some dirt from his cleats with the bat. His eyes met the pitcher's.

A sudden thought crossed Rhino's mind. Making an out here might not be so bad. The Mustangs would lose the game, but then they'd be heading home. He'd sleep in his own bed tonight.

That's not you, his thinker told him. *Be a winner.*

Rhino crushed the first pitch deep into right-center field. The ball smashed off the fence, and Rhino raced into third base with a stand-up triple.

Cooper scored, then Bella.

Rhino watched the scoreboard change: CARDINALS 1, MUSTANGS 2.

And that's how it looked when the game ended.

"A great defensive effort by both teams," said the announcer. "The Mustangs advance to tomorrow's championship game, right here at noon."

Two more teams began warming up on the field for the second game. The Mustangs bought lunch at the refreshment stand, then took seats in the bleachers to watch.

The team in black and orange had a strong pitcher. He fired a two-hit shutout.

"Glad we don't have to face him tomorrow," Cooper said. "Wonder who else they have?"

"We'll find out soon enough," Rhino said. He gave Cooper a light punch on the arm. Cooper would be pitching tomorrow.

That evening, a buffet was set up for the Mustangs at the hotel.

"I'm starving," Rhino said to Cooper as they waited in the line. He could see salads, roasted

chicken, macaroni and cheese, and other items, plus a dessert table with cookies, pies, and cake.

Rhino loaded up his plate. Winning always gave him a lift, especially when he drove in the winning runs. The pressure was high in this tournament, but he was having fun.

Grandpa phoned again while they were at the dinner table. Cooper's dad handed Rhino the phone.

Rhino told Grandpa all about the close game and his two hits. "He was the fastest pitcher I've ever faced," Rhino said. "But I figured him out."

Grandpa said they were on their way to C.J.'s science fair. Rhino loved doing things like that with his family. Why couldn't he be there?

"I miss you," Rhino said. His eyes stung again. Rhino walked away from his teammates. "Wish I was there with you."

"Tomorrow night will get here quickly," Grandpa said. "You keep your mind on the game. Have fun!"

Rhino waited a few minutes before returning to the table. Cooper was telling a joke about a Martian. Everybody was loose and happy.

Rhino sat quietly. He pushed some mac and cheese around his plate, but he wasn't hungry anymore.

Bella gently grabbed Rhino's arm. "Let's check out those cookies," she said.

The cookies did look good. When they were away from the others, Bella said, "Everything okay at home?"

"Sure. Everyone is fine."

"That's good."

"Wish I was with my family," Rhino said.

"I know what you mean," Bella replied. "But we're having lots of fun, aren't we? We can play some more games again in the room tonight. It'll be fun. You'll see."

You don't *know what I mean,* Rhino thought. *Your dad is here with you.*

But then his thinker told him to listen to his friend. *She's just trying to help.*

Rhino took three chocolate-chip cookies. "Thanks," he said softly to Bella.

But he didn't feel much better.

· CHAPTER 8 ·
Shedding the Blues

Cooper's dad watched a baseball game on TV while Rhino and Cooper played cards. Rhino tried to keep his mind off being homesick, but it wasn't working very well.

When Bella knocked on the door, Cooper hopped up.

"My dad said we can all hang out in our room tonight," Bella said. "We ordered ice cream from room service! Come on over."

"I'm on my way!" Cooper said. "Hurry up, Rhino."

Rhino set down the cards. "Go on ahead," he said. "I'll be along in a few minutes."

"If you're not, I'll be back to get you," Cooper said.

Rhino sat on the floor and watched the game.

"Tired?" Cooper's dad asked.

"Not really." Rhino stood up and looked out the window. He could see the stadium, which had a few lights on in the bleachers. The field was dark. For several minutes he watched cars and trucks go by on the street below.

He didn't feel like being with his teammates. They would just try to cheer him up. He wanted to make it through this night, play the game, and get home to Grandpa and C.J. as fast as possible.

But when Cooper came back for him, Rhino grabbed his Mustangs cap and headed over to Bella's room.

"Where have you been?" Carlos said. "The ice cream's getting soft."

Rhino smiled a little. "I like it that way," he said. "And you've got chocolate sauce on your nose, Carlos."

"We need some music," Bella said.

Rhino pointed out that the radio was already on.

"Live music," Bella said. "Carlos, wipe your nose clean and give us a show."

Carlos blushed. He'd won the school talent show with a soulful singing performance a few weeks before. Rhino and Cooper backed him up on guitar and drums.

Cooper tapped out a rhythm on the table with his fingertips. "Come on, Carlos," he said. "Belt one out!"

"Only if everybody helps me," Carlos said. "Listen. I'll teach you the words. You probably already know them. This one's easy."

Carlos turned his baseball cap around so the brim was behind him. He started to sing.

"Take me out to the ball game, take me out with the crowd . . ."

Bella and Cooper joined in. "Buy me some peanuts and Cracker Jack, I don't care if I never get back."

Soon everyone was singing. Rhino knew all the words, too. He jumped in for the last lines. "For it's one, two, three strikes you're out, at the old ball game!"

Everybody clapped. "Faster now!" Carlos said.

"Wait, wait!" said Rhino. "That last line? Isn't it 'one, two, three *stripes* you're out'?"

Carlos pulled off his hat and whacked Rhino with it.

"Hey," Rhino said, "that was your bad joke, Mr. Comedian!"

They sang the song several more times, getting faster all the time.

"Go, Mustangs!" Bella called as they finished.

Rhino looked around at his teammates. They did make him feel better, and they weren't even trying to. But someone was missing. "Where's Dylan?" he asked.

"He stayed back in the room," Carlos said. "Tired again, I guess."

Rhino didn't buy it. "Which room?" he asked.

"817."

Rhino said he'd be right back.

Dylan was wearing sunglasses when he answered the door.

"What's up?" Rhino asked.

Dylan touched the glasses but didn't take them off. "My eyes were . . . sore. Maybe from the glare of the sun this afternoon."

Rhino walked into the room. It was uncomfortable to be here with Dylan, since they'd never gotten along very well. But Rhino knew Dylan was upset. The party had made Rhino feel much better, and he wanted the same for Dylan.

"Everybody's having fun in Bella's room," Rhino said. "There's ice cream. Very mushy ice cream, but it's there."

"I'm . . . I'm fine here," Dylan said. "I was just about to go to bed."

But Dylan was fully dressed. Rhino pulled a chair from the desk and sat down. "You pitched an awesome game today."

"That's what I do," Dylan said. He stared at the wall. "I'll be at shortstop tomorrow, so . . ."

"Cooper will pitch fine."

"Not as good as I did."

Rhino rolled his eyes. Dylan was a good athlete, but he wasn't very humble about it. Rhino tried something else to break the tension. "Do you have a picture of Bruiser?"

"Yeah. On my phone." Dylan showed him the chubby puppy.

"Beautiful!" Rhino said. "He's all brown—even his eyes and his nose."

Dylan laughed. "He has a pink tongue."

"Wonder how he slept last night."

Dylan frowned. "Not so well, I hear. My parents won't let him up on the bed. They stuck him in a bathroom and let him cry. That's probably best so he won't get spoiled, but . . ."

"It's hard."

Dylan sniffed. He wiped his eye quickly under the sunglasses.

"He'll be okay," Rhino said. "It's tough being away. First from his mother, now from you."

"Yeah." Dylan sighed. He sat on the bed and looked at the picture on his phone.

"One more night," Rhino said. *That's all we have to get through.*

"I'll be right back," Dylan mumbled. He went into the bathroom. Rhino heard the water running for a couple of minutes.

"Okay," Dylan said. "What flavor ice cream?"

"Vanilla. But there's chocolate sauce, too."

"Sounds like my kind of party," Dylan said. "Let's go." He put his sunglasses back on, but Rhino could tell he was happier.

It doesn't take much, Rhino's thinker told him. *Just a few kind words sometimes.*

· CHAPTER 9 ·
Breakfast Surprise

Rhino was feeling much better when he crept beneath the covers of the cot that evening. *One more night,* he thought. *When I wake up tomorrow, it will be just like any other day. Soon I'll be back with C.J. and Grandpa James!*

"Hey, Rhino," Cooper whispered from his bed.

"Yeah?"

"Think we'll win tomorrow?"

"Of course I do!" Rhino laughed. "I always expect to win."

"I guess the pressure's on me this time," Cooper said.

"The Mustangs win together or we lose together," Rhino said. "That's the truth."

"Still . . . a few bad pitches and we're in trouble."

"Relax," Rhino said. "You've got the whole team behind you."

Rhino was half asleep when Cooper whispered again. "Rhino!"

"Yeah?"

"Are you homesick?"

Rhino rolled over and waited a moment. "A little," he finally replied.

"You aren't the only one."

"I know," Rhino said. "But the one who surprised me most was Carlos. If anybody was going to be homesick, I thought it would be him. He's so shy and he gets nervous. But he's been the most relaxed one of all—joking around, leading the songs."

"You never know," Cooper said.

Rhino nodded off. The sun was shining through the window by the time he woke up in the

morning. Cooper and his dad were watching the news on TV.

"I'm hungry!" Rhino said. "We need fuel for the big game."

"Then get dressed and let's go," Cooper said. "I've been up for an hour!"

"You could have gotten me up."

"I figured you needed the sleep."

Carlos and Bella were finishing their breakfast when Rhino and Cooper entered the hotel restaurant.

"We'll stay and keep you company," Carlos said, eating his last bite of pancake.

Rhino ordered a waffle with bacon and a side of melon. Cooper started to order, then his mouth dropped open and he stared past Rhino.

"What?" Rhino asked.

Cooper pointed. Rhino turned and couldn't believe what he saw. He jumped out of his chair.

"Grandpa!" Rhino called, hugging his grandfather and C.J. "I can't believe you're here."

"We got up super early and drove nonstop," C.J. said. "Didn't want to miss a championship game!"

"Any room for us at the table?" Grandpa asked. "We didn't even eat breakfast."

"We'll make room," Rhino said. "This is the best surprise ever!"

Grandpa whispered in Rhino's ear. "I'm very proud of you. And it has nothing to do with baseball."

Rhino's eyes stung again, but he wasn't sad this time. "I'm proud, too," he said.

Rhino and Cooper told C.J. all about the stadium, with its fancy scoreboard and flags everywhere. Then C.J. went on about the science fair.

"Third prize," he said. "Not bad, huh?"

"Excellent," Rhino said. He'd suggested that C.J. do a project about Jupiter's moons, since outer space was Rhino's favorite topic after dinosaurs. Outer space seemed better for a seventh-grade project.

"First prize was a robot made out of drinking straws and soda cans," C.J. said. "Really advanced. It had a claw to pick things up with."

"Wish I could have seen that," Rhino said.

"It was pretty cool," C.J. said. "I was happy with third place after I saw that one."

"First place is our target today," Rhino said. "That team we're up against looked really strong yesterday."

"No problem," Bella said. "We have the best power hitter in the tournament."

Rhino took a sip of water. He smiled at Bella, but he didn't want to brag about that game-winning triple.

When they reached the field, Rhino put the infielders through some quick drills, rolling fast grounders to them. "Accurate throws!" he called. *And square up,* his thinker reminded him.

The bleachers filled with spectators. Grandpa and C.J. had seats directly behind first base. Pop

music played from the announcer's booth. Rhino inhaled the smell of hot dogs cooking in the refreshment stand. The breeze was very light, so the pennants along the fences barely moved.

Rhino scuffed up the dirt near first base with his cleats. The orange-and-black-clad opponents were warming up in the outfield. The Chargers. They'd bat first today.

"It's funny that we're the home team," Rhino said to Carlos. "Since we're so far from home."

"Somebody had to be," Carlos replied. "I think the Chargers came even farther than we did."

Rhino noticed the pitcher from yesterday's game fielding pop flies. A different pitcher was throwing to the catcher on the other side of third base. He was tall and thin, with a strong overhand motion.

This would be a tough game. So many people watching; so far from home. But Rhino didn't feel nervous at all. *My kind of day,* he thought. He turned and gave a thumbs-up to C.J. and Grandpa.

The Mustangs gathered in the dugout. Coach Ray read off the positions and batting order:

1) Cooper P
2) Bella   RF
3) Dylan   SS
4) Rhino   1B
5) Manny   LF
6) Carlos 2B
7) Sara   3B
8) Paul   CF
9) Gabe   C

"Hands in," Coach said.

"One . . . two . . . three . . . Mustangs!"

Everyone cheered as the Mustangs trotted to their positions. Cooper fired some warm-up pitches. Rhino tossed the ball around the horn.

The curly-haired Charger who pitched yesterday swung a bat near home plate, watching Cooper. He'd be leading off.

Rhino took a deep breath.

"Batter up!" called the umpire.

Play ball, Rhino thought.

· CHAPTER 10 ·
A Slugfest

It was *not* a good start.

Rhino stood tensed, a few feet from first base, ready to dodge either way to field the ball or race to the bag for a throw.

Cooper went into his windup. The pitch was a blur—fast and straight.

Thwack!

Rhino knew as soon as he heard it. *Good-bye, baseball!*

The ball streaked over the left field fence on a line drive. One pitch. One run.

Rhino nodded to the kid as he rounded first base. Then he jogged to the mound.

"No problem," Rhino said, patting Cooper's shoulder. "You got the bad pitch out of the way. Now shut them down."

Cooper wiped his forehead with his glove and kicked at the dirt. "It was a good pitch," he muttered.

"Good for *him*," Rhino said with a smirk. He trotted back to first base.

"Okay, Cooper, let's go!" shouted Bella.

"You're the man!" called Carlos. "We've got you covered."

But Cooper was rattled. He walked the next two batters. Coach Ray called for time and walked to the mound.

Rhino came over again, too.

"Feel okay?" Coach asked.

Cooper nodded. "I do."

"Just find the plate," Rhino said. "Forget about

that home run. He's their best player. These other guys are just regular."

Cooper blew his breath out hard. "I'm okay," he said again.

He struck out the next two batters, then gave up a single that scored a second run.

"Let's get those runs back!" Rhino said when the Mustangs went to the dugout. "Starting with you, Coop."

The Chargers' pitcher was hard to read. He threw a mix of fastballs and sinkers, with a few curveballs, too. Cooper and Bella struck out before Dylan was back at the plate again to launch a fly ball that bounced off the left-field fence. He cruised into second easily.

"Bring him in, Rhino," Carlos yelled.

Fastball, low and inside.

"Strike one!"

Another fastball, even lower.

"Ball!"

The next pitch came in slower, and it dropped before the plate. It hopped past the catcher, and Rhino stepped back. Dylan ran toward third and dove ahead of the throw.

"Safe!"

Rhino was ahead in the count: two balls and a strike. *This is the one,* he thought.

He never lost sight of the next pitch, and met it with a powerful swing. The ball rolled straight up the middle for a clean single. Dylan scored. The Mustangs were back in the game.

But the Chargers scored two more in the second and another in the third. This game was nothing like the day before. After Rhino doubled to drive in Cooper and Bella, he scored on Carlos's single. That made it 5–4, Chargers.

"That's a lot of runs in three innings," Cooper said as they took the field for the fourth.

"It's that kind of game," Rhino said. "We're right in it."

The runs kept coming. Two for the Chargers in the top of the fourth, two more for the Mustangs in the bottom.

Rhino had never been in such a high-scoring game. It stood at 7–6 when the Mustangs came to bat in the bottom of the sixth.

"Last ups!" Rhino said. "It's now or never."

It's been a long time since we lost a game, Rhino thought. The Mustangs had seven wins in a row. And they'd come from behind several times. Why should today be any different? But somebody needed to get on base. Otherwise Rhino's day was over.

Cooper lined the first pitch up the middle for a single. Rhino picked up his bat. "Let's go, Bella!" he called.

Bella tapped a perfect bunt up the third baseline. The throw to first beat her by half a step, but Cooper slid safely into second.

"Come on, Dylan," Rhino said.

Dylan patiently watched two pitches go by.

"Good eye!" Carlos called.

Dylan swung hard at the next one, but he got under the ball. It soared high into the air, but not very deep. The second baseman easily made the catch.

Rhino stepped to the plate with two outs.

"Tie this game up!" Bella called. "Show them that power!"

Rhino had already hit a single and a double today to go with his single and triple from yesterday. Only one thing was missing from his list: a home run.

He fouled the first pitch down the first baseline, where it crashed into an empty section of the bleachers.

"Straighten it out!" came a call from the dugout.

Rhino swung late on the second pitch and sent it deep to the left, but foul.

The Mustangs were down to their last strike.

Just meet the ball, Rhino's thinker told him. *Get Cooper home and get on base!*

The pitch was low. A little inside. But too close to the strike zone to let it go by.

Rhino swung. The ball zinged hard between first base and second and dropped into right field. Cooper sprinted around third and never let up. He slid into home, barely beating the throw. Rhino dove into second.

All the Mustangs were standing and cheering. Rhino brushed the dirt from his jersey. He was in position to score the tying run.

With first base empty, the Chargers' pitcher didn't take any chances. Manny walked on five pitches.

"Any base!" called the shortstop—the guy who'd led off the game with a home run. The Chargers could get a force out at first, second, or third on a groundball.

It was all up to Carlos. He didn't have much power, but he'd singled twice today. Another one would bring Rhino home.

Now the Mustangs chanted a name they'd never chanted before: "Carlos. Carlos. Carlos."

Carlos watched a strike go by. He leaned back for the next one. Then he hit the ball harder than he'd ever hit it in his life. Rhino sprinted as soon as he heard the *Whack!*

Rhino was at full speed, grinding toward third, when a huge yell went up in the bleachers. The third baseman jumped with both hands in the air, and the other Chargers ran toward the mound.

Rhino looked back. The shortstop was holding up his glove. He reached in and pulled out the ball. Carlos had lined the ball right at him. Game over.

Rhino grabbed Carlos and patted his back. "Great game," he said. "Great tournament."

"You too."

As the teams lined up to shake hands, Rhino spent an extra few seconds with the shortstop.

"You made the difference," Rhino said. "Hitting and fielding."

"You guys are the best team we've ever played," the kid replied. "Congratulations. Hope we'll see you again someday."

The Mustangs were upbeat as they loaded onto the bus. Nobody liked to lose, but they knew the game could have gone either way.

Rhino shook hands with Bella, Cooper, and the others. "I'm riding home with my family," he said. "I'll see you all in school tomorrow."

"Go, Mustangs!" Bella said.

It was a beautiful afternoon. Grandpa kept the car windows open a bit and they listened to music on the radio.

"Three hits in one game is pretty impressive, little brother," C.J. said.

"Yeah, I had a good tournament. Too bad we lost, but it was a lot of fun."

"We'll celebrate with a special dinner tonight," Grandpa said. "Anything you want."

"Sounds great," Rhino replied. But dinner was

a long way off. He was already hungry. He was also very happy.

Rhino remembered something. He dug into his knapsack and pulled out the second PB&J sandwich he'd made on Friday.

"Isn't that a little stale?" C.J. asked as Rhino unwrapped it.

Rhino shook his head. He gobbled every bite. The sandwich couldn't have tasted better.

It reminded him of home.

THE END OF A PERFECT SEASON!

HERE'S A SNEAK PEEK AT LITTLE RHINO #6: *TROPHY NIGHT!*

CHAPTER ONE
The Trophy Table

Rhino ran his fingers over the letters on the trophy: MVP.

Most Valuable Player. The shiny metal baseball player was frozen in mid-swing at the top of the trophy. *Smacking a home run,* Rhino thought.

Rhino had hit quite a few homers for the Mustangs. He'd been having a great baseball season. But the trophy wasn't his. It belonged to his brother, C.J.from two years ago.

Rhino's first baseball season was nearing its end. He hoped he'd soon be bringing home a trophy like C.J.'s.

MVP of this league. Then record-setting home-run hitter in middle school. Player of the Year in high school and college. Then the Major Leagues. The all-star game! The World Series!

Rhino admired the trophy again, feeling the smooth metal.

"Making room for my new one?" C.J. said with a laugh as he entered the living room. Rhino and his older brother looked a lot alike, but C.J. was taller and more muscular. They both had a quick smile.

"Or mine," Rhino said. He set the trophy back on the ledge, between C.J.'s championship basketball and soccer trophies.

Grandpa James had set up a shelf in the living room and filled it with all of their accolades. Rhino spent a lot of time looking at the various trophies and plaques they had collected as a family.

"Here's the most important one yet," said Grandpa James, following C.J. into the room. He held C.J.'s latest award—a third-place plaque from

the middle-school science fair. Sports were very important in Grandpa James's house, but school and learning always came first. C.J.'s prize-winning poster about Jupiter's moons had taken a lot of brainwork. He'd stayed up late several nights researching the facts, then carefully drew the moons circling the giant planet.

"That's quite a lineup of awards," Grandpa said, placing his hand on Rhino's shoulder. "And there will be one for you soon. Every player in your baseball league gets a trophy for taking part."

Rhino nodded, but just "taking part" wasn't enough for him. He wanted a trophy for best hitter, or all-star first baseman, or most home runs. And another one for winning the championship. The Mustangs had hit a rough stretch and lost their two most recent games, but they could wrap up a spot in the playoffs by winning their final regular-season game this weekend. From there, they'd have a shot at the title.

C.J. pointed to an older trophy behind the

others. "There's Grandpa's league championship award from high school basketball," he said. "That's a big-time trophy."

"We won that game on a last-second shot," Grandpa said. "I grabbed a rebound, gave a quick fake, then dished the ball to my teammate. He scored at the buzzer."

Everybody in this family has won major sports awards, Rhino thought. *Except me.* His only contribution to the table was a blue ribbon from the school talent show. He'd performed with two of his teammates and won first prize, but most of that credit belonged to Carlos, the singer in the Mustang Rock band. Rhino had helped Carlos gain enough confidence to use his great singing voice in the talent show.

Rhino's thinker told him not to worry. He'd earn a sports trophy soon.

"Little Rhino, you should hear more about how C.J. earned that MVP trophy," Grandpa said.

"For being the star, right?" Rhino asked.

Grandpa shrugged. "He was a star, but . . . were you the best hitter on that team, C.J.?

C.J. shook his head. "Bobby had a higher batting average."

"And you didn't hit the most home runs did you?"

C.J. laughed. "I didn't hit *any* home runs that season. I wasn't as strong then as Rhino is now."

Rhino found that hard to believe. C.J. was so strong. He'd hit a lot of homers this year for his middle-school team.

"Seems to me you weren't the star pitcher either," Grandpa said. "But you did many things well, C.J. The best thing you did was support your teammates. You were the team leader. You always had a good word for everybody, whether they hit a home run or struck out."

I do that, too, Rhino thought. *Maybe I am an MVP.*

"Of course, he was an excellent player, too," Grandpa said. "But there a lot of things that go into being an MVP."

They celebrated C.J.'s science award with Grandpa's spaghetti and meatballs, then ice cream for dessert. Rhino scooped out the chocolate-chip ice cream and drizzled it with chocolate syrup. He added a fresh strawberry to the top of the pile. *It looks a little like a trophy,* Rhino thought. He imagined his coach handing him a huge baseball trophy— just like C.J.'s but a foot taller. The lettering said PLAYER OF THE YEAR.

"What are you waiting for?" C.J. asked. "Your dessert will melt."

Rhino nodded. He took a big spoonful and grinned. "I was just thinking," he said. "Maybe soon we'll have a celebration dinner for me."

CHAPTER TWO
Bases Loaded

Rhino smacked his hand into his glove and eyed the runner at first base. "Let's turn two!" he called to the infielders. A double play would end the inning and maintain the Mustangs' one-run lead.

It had been a wild game, with the Wolves taking a 2-0 lead back in the first inning before there were any outs. But Dylan had settled down for the Mustangs and struck out three batters in a row. Then the Mustangs stormed back with four runs in the bottom of the first inning, including an RBI double by Rhino.

The lead seesawed back and forth for several innings. In the top of the fifth, Rhino led off with a walk, stole two bases, and scored from third on a wild pitch. That gave the Mustangs a 7-6 lead.

"Come on, Dylan!" called Rhino's teammates. "Strike this guy out!"

The batter sent a bouncing rocket up the middle, skimming past Dylan's glove. Cooper darted over from shortstop and snagged the ball. He flipped it to Carlos at second base for an out.

Rhino's eyes widened as Carlos's throw to first soared too high. With the batter storming toward the base, Rhino leaped, knocking the ball down with the tip of his glove. The ball skipped into right field, but Rhino ran after it, and the runner did not try for second.

"Two outs," Rhino called as he flipped the ball to Dylan. He held up two fingers and waved to his teammates.

The day was hot. Rhino wiped his sweaty forehead with the back of his hand and pulled his cap

into place. He wanted a drink of water. "Let's end this inning!" he yelled.

Falling behind this late in the game would mean trouble. If the Mustangs lost, their season would be over.

I like baseball too much to let that happen, Rhino thought. *I wish we could play all summer!*

Rhino had studied the league standings before the game. With a win, the Mustangs' final record would be eight wins and four losses. That would put them in a tie for second place. But a loss today would drop their record to 7-5—in a tie for fourth with the Wolves. Since only four teams would make the playoffs, the final spot would go to the Wolves based on today's result.

The runner at first base took a big lead. Rhino knew that the kid was fast. He could score from first on a double.

Dylan wound up and threw a fastball.

Whack! The ball sped toward right field. Rhino dove, leaving his feet and stretching as far as he

could go. He grabbed the line drive, rolling in the grass and squeezing the ball tight. He held up his glove.

"Out!" called the umpire.

Rhino smiled. His teammates cheered as they ran to the dugout.

"Super catch!" said Bella, running in from right field. She grabbed Rhino's hand and helped him to his feet. "That ball would have reached the fence. He hit it so hard it might have smashed through it!"

The Mustangs had the top of their batting order coming up: Cooper, Bella, and Dylan. Rhino would bat fourth—the cleanup hitter—as long as someone got on base.

"Big rally!" Rhino said. "Let's wrap this up." He was puffing from excitement. Making that catch had been a thrill.

Rhino took a long swig of water. "How's your arm?" he asked Dylan. "You've thrown a lot of pitches today."

"I'm fine," Dylan replied, staring out at the field. Dylan was never very friendly, but he and Rhino had grown to respect each other as athletes. "The hot weather keeps me loose."

Cooper popped the ball into left field behind the shortstop. It looked like an easy out. But the shortstop and the left fielder both ran toward it, then stopped when they saw each other. The ball fell softly to the grass between them, and Cooper reached first base with a single.

Bella made a perfect bunt down the third-base line, moving Cooper to second. Better yet, she beat the catcher's throw to first.

"Way to hustle, Bella!" Rhino called.

Two on, no outs.

"Bring them home, Dylan," Rhino said, stepping into the on-deck circle. He took a few easy swings.

The Wolves' coach called timeout and walked to the mound. He spoke to the pitcher for a minute, then waved to the third baseman. "Pitching change,"

the coach called to the umpire. The third baseman and the pitcher switched positions.

Dylan walked over to Rhino as the new pitcher warmed up. "This guy's in my class at school," Dylan whispered. "He's even wilder than that first pitcher. If he can't find the strike zone, I'll get a walk."

Rhino nodded. Dylan's guess was right. The pitcher managed just one strike.

"Ball four!" the umpire called. "Take your base."

Dylan trotted to first, and Cooper and Bella moved up. The bases were loaded.

LET THE GAME BEGIN!

YOU CAN'T HIT WITHOUT A BAT!

BATTER UP WITH
LITTLE RHINO #2!